EXCERPTS OF CHAPTERS 10 & 11 FROM

Application-Specific Integrated Circuits

AN INTRODUCTION TO VHDL & VERILOG HDL

Michael John Sebastian Smith

Boston San Francisco New York
London Toronto Sydney Tokyo Singapore Madrid
Mexico City Munich Paris Cape Town Hong Kong Montreal

VHDL

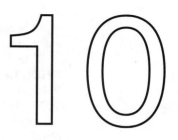

The U.S. Department of Defense (DoD) supported the development of **VHDL** (**VHSIC hardware description language**) as part of the **VHSIC** (**very high-speed IC**) program in the early 1980s. The companies in the VHSIC program found they needed something more than schematic entry to describe large ASICs, and proposed the creation of a hardware description language. VHDL was then handed over to the Institute of Electrical and Electronics Engineers (IEEE) in order to develop and approve the IEEE Standard 1076-1987.[1] As part of its standardization process the DoD has specified the use of VHDL as the documentation, simulation, and verification medium for ASICs (MIL-STD-454). Partly for this reason VHDL has gained

---

[1]Some of the material in this chapter is reprinted with permission from IEEE Std 1076-1993, © 1993 IEEE. All rights reserved.

rapid acceptance, initially for description and documentation, and then for design entry, simulation, and synthesis as well.

The first revision of the 1076 standard was approved in 1993. References to the ·VHDL **Language Reference Manual (LRM)** in this chapter—[VHDL 87LRM2.1, 93LRM2.2] for example—point to the 1987 and 1993 versions of the LRM [IEEE, 1076-1987 and 1076-1993]. The prefixes 87 and 93 are omitted if the references are the same in both editions. Technically 1076-1987 (known as VHDL-87) is now obsolete and replaced by 1076-1993 (known as VHDL-93). Except for code that is marked 'VHDL-93 only' the examples in this chapter can be **analyzed** (the VHDL word for "compiled") and simulated using both VHDL-87 and VHDL-93 systems.

10.1   A Counter

The following VHDL model describes an electrical "black box" that contains a 50 MHz clock generator and a counter. The counter increments on the negative edge of the clock, counting from zero to seven, and then begins at zero again. The model contains separate *processes* that execute at the same time as each other. Modeling concurrent execution is the major difference between HDLs and computer programming languages such as C.

```
entity Counter_1 is end; -- declare a "black box" called Counter_1
library STD; use STD.TEXTIO.all; -- we need this library to print
architecture Behave_1 of Counter_1 is -- describe the "black box"
-- declare a signal for the clock, type BIT, initial value '0'
    signal Clock : BIT := '0';
-- declare a signal for the count, type INTEGER, initial value 0
    signal Count : INTEGER := 0;
begin
    process begin -- process to generate the clock
        wait for 10 ns; -- a delay of 10 ns is half the clock cycle
        Clock <= not Clock;
        if (now > 340 ns) then wait; end if; -- stop after 340 ns
    end process;
-- process to do the counting, runs concurrently with other processes
    process begin
-- wait here until the clock goes from 1 to 0
        wait until (Clock = '0');
-- now handle the counting
        if (Count = 7) then Count <= 0;
        else Count <= Count + 1;
        end if;
    end process;
    process (Count) variable L: LINE; begin -- process to print
        write(L, now); write(L, STRING'(" Count="));
        write(L, Count); writeline(output, L);
    end process;
end;
```

Throughout this book VHDL **keywords** (reserved words that are part of the language) are shown in bold type in code examples (but not in the text). The code examples use the bold keywords to improve readability. VHDL code is often lengthy and the code in this book is always complete wherever possible. In order to save space many of the code examples do not use the conventional spacing and formatting that is normally considered good practice. So "Do as I say and not as I do."

The steps to simulate the model and the printed results for `Counter_1` using the Model Technology V-System/Plus common-kernel simulator are as follows:

```
> vlib work
> vcom Counter_1.vhd
Model Technology VCOM V-System VHDL/Verilog 4.5b
-- Loading package standard
-- Compiling entity counter_1
-- Loading package textio
-- Compiling architecture behave_1 of counter_1
> vsim -c counter_1
# Loading /../std.standard
# Loading /../std.textio(body)
# Loading work.counter_1(behave_1)
VSIM 1> run 500
# 0 ns Count=0
# 20 ns Count=1
(...15 lines omitted...)
# 340 ns Count=1
VSIM 2> quit
>
```

10.2   A 4-bit Multiplier

This section presents a more complex VHDL example to motivate the study of the syntax and semantics of VHDL in the rest of this chapter.

10.2.1   An 8-bit Adder

Table 10.1 shows a VHDL model for the full adder that we described in Section 2.6, "Datapath Logic Cells." Table 10.2 shows a VHDL model for an 8-bit ripple-carry adder that uses eight instances of the full adder.

10.2.2   A Register Accumulator

Table 10.3 shows a VHDL model for a positive-edge–triggered D flip-flop with an active-high asynchronous clear. Table 10.4 shows an 8-bit register that uses this D flip-flop model (this model only provides the Q output from the register and leaves the QN flip-flop outputs unconnected).

TABLE 10.1   A full adder.

```
entity Full_Adder is                                          --1
   generic (TS : TIME := 0.11 ns; TC : TIME := 0.1 ns);       --2
   port (X, Y, Cin: in BIT; Cout, Sum: out  BIT);             --3
end Full_Adder;                                               --4

architecture Behave of Full_Adder is                          --5
begin                                                         --6
Sum  <= X xor Y xor Cin after TS;                             --7
Cout <= (X and Y) or (X and Cin) or (Y and Cin) after TC;     --8
end;                                                          --9
```

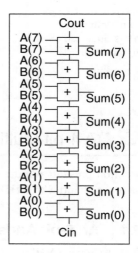

Timing:
TS (Input to Sum) = 0.11 ns
TC (Input to Cout) = 0.1 ns

TABLE 10.2   An 8-bit ripple-carry adder.

```
entity Adder8 is                                              --1
   port (A, B: in BIT_VECTOR(7 downto 0);                     --2
   Cin: in BIT; Cout: out BIT;                                --3
   Sum: out BIT_VECTOR(7 downto 0));                          --4
end Adder8;                                                   --5

architecture Structure of Adder8 is                           --6
component Full_Adder                                          --7
port (X, Y, Cin: in BIT; Cout, Sum: out BIT);                 --8
end component;                                                --9
signal C: BIT_VECTOR(7 downto 0);                             --10
begin                                                         --11
Stages: for i in 7 downto 0 generate                          --12
   LowBit: if i = 0 generate                                  --13
   FA:Full_Adder port map (A(0),B(0),Cin,C(0),Sum(0));        --14
   end generate;                                              --15
   OtherBits: if i /= 0 generate                              --16
   FA:Full_Adder port map                                     --17
      (A(i),B(i),C(i-1),C(i),Sum(i));                         --18
   end generate;                                              --19
end generate;                                                 --20
Cout <= C(7);                                                 --21
end;                                                          --22
```

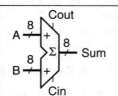

TABLE 10.3   Positive-edge–triggered D flip-flop with asynchronous clear.

```
entity DFFClr is                                       --1
   generic(TRQ : TIME := 2 ns; TCQ : TIME := 2 ns);    --2
   port (CLR, CLK, D : in BIT; Q, QB : out BIT);       --3
end;                                                   --4

architecture Behave of DFFClr is                       --5
signal Qi : BIT;                                       --6
begin QB <= not Qi; Q <= Qi;                           --7
process (CLR, CLK) begin                               --8
   if CLR = '1' then Qi <= '0' after TRQ;              --9
   elsif CLK'EVENT and CLK = '1'                       --10
      then Qi <= D after TCQ;                          --11
   end if;                                             --12
end process;                                           --13
end;                                                   --14
```

Timing:
TRQ (CLR to Q/QN) = 2 ns
TCQ (CLK to Q/QN) = 2 ns

TABLE 10.4   An 8-bit register.

```
entity Register8 is                                    --1
   port (D : in BIT_VECTOR(7 downto 0);                --2
   Clk, Clr: in BIT ; Q : out BIT_VECTOR(7 downto 0)); --3
end;                                                   --4

architecture Structure of Register8 is                 --5
   component DFFClr                                     --6
      port (Clr, Clk, D : in BIT; Q, QB : out BIT);     --7
   end component;                                       --8
   begin                                                --9
      STAGES: for i in 7 downto 0 generate              --10
      FF: DFFClr port map (Clr, Clk, D(i), Q(i), open); --11
      end generate;                                     --12
end;                                                   --13
```

8-bit register. Uses DFFClr positive edge-triggered flip-flop model.

Table 10.5 shows a model for a datapath multiplexer that consists of eight 2:1 multiplexers with a common select input (this select signal would normally be a control signal in a datapath). The multiplier will use the register and multiplexer components to implement a register accumulator.

10.2.3   Zero Detector

Table 10.6 shows a model for a variable-width zero detector that accepts a bus of any width and will produce a single-bit output of '1' if all input bits are zero.

TABLE 10.5   An 8-bit multiplexer.

```
entity Mux8 is                                            --1
  generic (TPD : TIME := 1 ns);                           --2
  port (A, B : in BIT_VECTOR (7 downto 0);                --3
    Sel : in BIT := '0'; Y : out BIT_VECTOR (7 downto 0));  --4
end;                                                      --5

architecture Behave of Mux8 is                            --6
begin                                                     --7
    Y <= A after TPD when Sel = '1' else B after TPD;     --8
end;                                                      --9
```

Eight 2:1 MUXs with single select input.
Timing:
TPD(input to Y)=1 ns

TABLE 10.6   A zero detector.

```
entity AllZero is                                         --1
  generic (TPD : TIME := 1 ns);                           --2
  port (X : BIT_VECTOR; F : out BIT );                    --3
end;                                                      --4

architecture Behave of AllZero is                         --5
begin process (X) begin F <= '1' after TPD;               --6
  for j in X'RANGE loop                                   --7
    if X(j) = '1' then F <= '0' after TPD; end if;        --8
  end loop;                                               --9
end process;                                              --10
end;                                                      --11
```

Variable-width zero detector.
Timing:
TPD(X to F) = 1 ns

10.2.4   A Shift Register

Table 10.7 shows a variable-width shift register that shifts (left or right under input control, DIR) on the positive edge of the clock, CLK, gated by a shift enable, SH. The parallel load, LD, is synchronous and aligns the input LSB to the LSB of the output, filling unused MSBs with zero. Bits vacated during shifts are zero filled. The clear, CLR, is asynchronous.

10.2.5   A State Machine

To multiply two binary numbers A and B, we can use the following algorithm:

1.   If the LSB of A is '1', then add B into an accumulator.

2.   Shift A one bit to the right and B one bit to the left.

3.   Stop when all bits of A are zero.

TABLE 10.7   A variable-width shift register.

```
entity ShiftN is                                           --1
  generic (TCQ : TIME := 0.3 ns; TLQ : TIME := 0.5 ns;     --2
    TSQ : TIME := 0.7 ns);                                 --3
  port(CLK, CLR, LD, SH, DIR: in BIT;                      --4
    D: in BIT_VECTOR; Q: out BIT_VECTOR);                  --5
  begin assert (D'LENGTH <= Q'LENGTH)                      --6
    report "D wider than output Q" severity Failure;       --7
end ShiftN;                                                --8

architecture Behave of ShiftN is                           --9
  begin Shift: process (CLR, CLK)                          --10
  subtype InB  is NATURAL range D'LENGTH-1 downto 0;       --11
  subtype OutB is NATURAL range Q'LENGTH-1 downto 0;       --12
  variable St: BIT_VECTOR(OutB);                           --13
  begin                                                    --14
    if CLR = '1' then                                      --15
      St := (others => '0'); Q <= St after TCQ;            --16
    elsif CLK'EVENT and CLK='1' then                       --17
      if LD = '1' then                                     --18
        St := (others => '0');                             --19
        St(InB) := D;                                      --20
        Q <= St after TLQ;                                 --21
      elsif SH = '1' then                                  --22
        case DIR is                                        --23
        when '0' => St := '0' & St(St'LEFT downto 1);      --24
        when '1' => St := St(St'LEFT-1 downto 0) & '0';    --25
        end case;                                          --26
        Q <= St after TSQ;                                 --27
      end if;                                              --28
    end if;                                                --29
  end process;                                             --30
end;                                                       --31
```

CLK	Clock
CLR	Clear, active high
LD	Load, active high
SH	Shift, active high
DIR	Direction, 1 = left
D	Data in
Q	Data out

Variable-width shift register. Input width must be less than output width. Output is left-shifted or right-shifted under control of DIR. Unused MSBs are zero-padded during load. Clear is asynchronous. Load is synchronous.

Timing:
TCQ (CLR to Q) = 0.3ns
TLQ (LD to Q) = 0.5ns
TSQ (SH to Q) = 0. 7ns

Table 10.8 shows the VHDL model for a Moore (outputs depend only on the state) finite-state machine for the multiplier, together with its state diagram.

10.2.6   A Multiplier

Table 10.9 shows a schematic and the VHDL code that describes the interconnection of all the components for the multiplier. Notice that the schematic comprises two halves: an 8-bit-wide datapath section (consisting of the registers, adder, multiplexer,

TABLE 10.8   A Moore state machine for the multiplier.

```
entity SM_1 is                                  --1
  generic (TPD : TIME := 1 ns);                 --2
  port(Start, Clk, LSB, Stop, Reset: in BIT;    --3
  Init, Shift, Add, Done : out BIT);            --4
end;                                            --5

architecture Moore of SM_1 is                   --6
type STATETYPE is (I, C, A, S, E);              --7
signal State: STATETYPE;                        --8
begin                                           --9
Init <= '1' after TPD when State = I            --10
  else '0' after TPD;                           --11
Add  <= '1' after TPD when State = A            --12
  else '0' after TPD;                           --13
Shift <= '1' after TPD when State = S           --14
  else '0' after TPD;                           --15
Done <= '1' after TPD when State = E            --16
  else '0' after TPD;                           --17
process (CLK, Reset) begin                      --18
  if Reset = '1' then State <= E;               --19
  elsif CLK'EVENT and CLK = '1' then            --20
    case State is                               --21
    when I => State <= C;                       --22
    when C =>                                   --23
      if LSB = '1' then State <= A;             --24
      elsif Stop = '0' then State <= S;         --25
      else State <= E;                          --26
      end if;                                   --27
    when A => State <= S;                       --28
    when S => State <= C;                       --29
    when E =>                                   --30
      if Start = '1' then State <= I; end if;   --31
    end case;                                   --32
  end if;                                       --33
end process;                                    --34
end;                                            --35
```

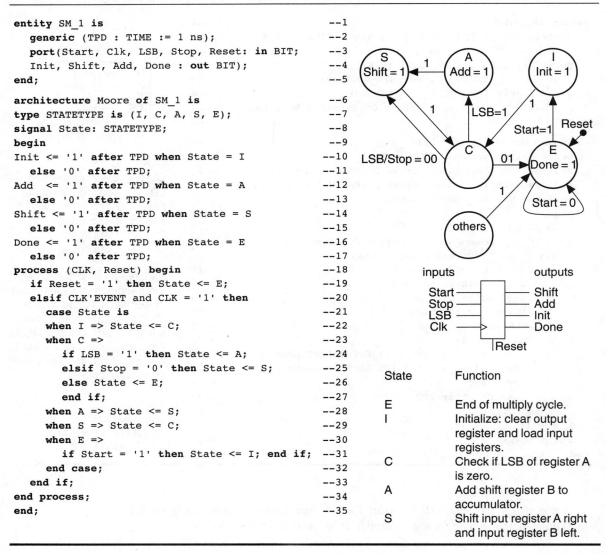

State	Function
E	End of multiply cycle.
I	Initialize: clear output register and load input registers.
C	Check if LSB of register A is zero.
A	Add shift register B to accumulator.
S	Shift input register A right and input register B left.

and zero detector) and a control section (the finite-state machine). The arrows in the schematic denote the inputs and outputs of each component. As we shall see in Section 10.7, VHDL has strict rules about the direction of connections.

TABLE 10.9   A 4-bit by 4-bit multiplier.

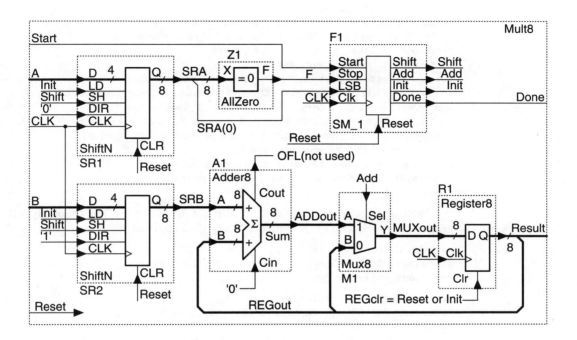

```
entity Mult8 is                                                                      --1
port (A, B: in BIT_VECTOR(3 downto 0); Start, CLK, Reset: in BIT;                    --2
Result: out BIT_VECTOR(7 downto 0); Done: out BIT); end Mult8;                       --3

architecture Structure of Mult8 is use work.Mult_Components.all;                     --4
signal SRA, SRB, ADDout, MUXout, REGout: BIT_VECTOR(7 downto 0);                     --5
signal Zero, Init, Shift, Add, Low: BIT := '0'; signal High: BIT := '1';             --6
signal F, OFL, REGclr: BIT;                                                          --7
begin                                                                                --8
REGclr <= Init or Reset; Result  <= REGout;                                          --9
SR1 : ShiftN   port map(CLK=>CLK,CLR=>Reset,LD=>Init,SH=>Shift,DIR=>Low ,D=>A,Q=>SRA); --10
SR2 : ShiftN   port map(CLK=>CLK,CLR=>Reset,LD=>Init,SH=>Shift,DIR=>High,D=>B,Q=>SRB); --11
Z1 : AllZero   port map(X=>SRA,F=>Zero);                                             --12
A1 : Adder8    port map(A=>SRB,B=>REGout,Cin=>Low,Cout=>OFL,Sum=>ADDout);            --13
M1 : Mux8      port map(A=>ADDout,B=>REGout,Sel=>Add,Y=>MUXout);                     --14
R1 : Register8 port map(D=>MUXout,Q=>REGout,Clk=>CLK,Clr=>REGclr);                   --15
F1 : SM_1      port map(Start,CLK,SRA(0),Zero,Reset,Init,Shift,Add,Done);            --16
end;                                                                                 --17
```

10.2.7   Packages and Testbench

To complete and test the multiplier design we need a few more items. First we need the following "components list" for the items in Table 10.9:

```
package Mult_Components is                                    --1
component Mux8 port (A,B:BIT_VECTOR(7 downto 0);             --2
   Sel:BIT;Y:out BIT_VECTOR(7 downto 0));end component;      --3
component AllZero port (X : BIT_VECTOR;                       --4
   F:out BIT );end component;                                 --5
component Adder8 port (A,B:BIT_VECTOR(7 downto 0);Cin:BIT;   --6
   Cout:out BIT;Sum:out BIT_VECTOR(7 downto 0));end component; --7
component Register8 port (D:BIT_VECTOR(7 downto 0);          --8
   Clk,Clr:BIT; Q:out BIT_VECTOR(7 downto 0));end component; --9
component ShiftN port (CLK,CLR,LD,SH,DIR:BIT;D:BIT_VECTOR;   --10
   Q:out BIT_VECTOR);end component;                          --11
component SM_1 port (Start,CLK,LSB,Stop,Reset:BIT;           --12
   Init,Shift,Add,Done:out BIT);end component;               --13
end;                                                          --14
```

Next we need some utility code to help test the multiplier. The following VHDL generates a clock with programmable "high" time (HT) and "low" time (LT):

```
package Clock_Utils is                                        --1
procedure Clock (signal C: out Bit; HT, LT:TIME);           --2
end Clock_Utils;                                             --3

package body Clock_Utils is                                  --4
procedure Clock (signal C: out Bit; HT, LT:TIME) is         --5
begin                                                        --6
   loop C<='1' after LT, '0' after LT + HT; wait for LT + HT; --7
   end loop;                                                 --8
end;                                                         --9
end Clock_Utils;                                             --10
```

Finally, the following code defines two functions that we shall also use for testing—the functions convert an array of bits to a number and vice versa:

```
package Utils is                                             --1
   function Convert (N,L: NATURAL) return BIT_VECTOR;       --2
   function Convert (B: BIT_VECTOR) return NATURAL;         --3
end Utils;                                                  --4

package body Utils is                                       --5
   function Convert (N,L: NATURAL) return BIT_VECTOR is     --6
      variable T:BIT_VECTOR(L-1 downto 0);                 --7
      variable V:NATURAL:= N;                               --8
      begin for i in T'RIGHT to T'LEFT loop                --9
         T(i) := BIT'VAL(V mod 2); V:= V/2;                --10
      end loop; return T;                                   --11
   end;                                                     --12
```

```
function Convert (B: BIT_VECTOR) return NATURAL is          --13
   variable T:BIT_VECTOR(B'LENGTH-1 downto 0) := B;          --14
   variable V:NATURAL:= 0;                                   --15
   begin for i in T'RIGHT to T'LEFT loop                     --16
      if T(i) = '1' then V:= V + (2**i); end if;             --17
      end loop; return V;                                    --18
   end;                                                      --19
end Utils;                                                   --20
```

The following code tests the multiplier model. This is a **testbench** (this simple
example is not a comprehensive test). First we reset the logic (line 17) and then
apply a series of values to the inputs, A and B. The clock generator (line 14) supplies
a clock with a 20 ns period. The inputs are changed 1 ns after a positive clock edge,
and remain stable for 20 ns through the next positive clock edge.

```
entity Test_Mult8_1 is end; -- runs forever, use break!!          --1
architecture Structure of Test_Mult8_1 is                         --2
use Work.Utils.all; use Work.Clock_Utils.all;                     --3
   component Mult8 port                                            --4
      (A, B : BIT_VECTOR(3 downto 0); Start, CLK, Reset : BIT;     --5
      Result : out BIT_VECTOR(7 downto 0); Done : out BIT);        --6
   end component;                                                  --7
signal A, B : BIT_VECTOR(3 downto 0);                             --8
signal Start, Done : BIT := '0';                                  --9
signal CLK, Reset : BIT;                                          --10
signal Result : BIT_VECTOR(7 downto 0);                           --11
signal DA, DB, DR : INTEGER range 0 to 255;                       --12
begin                                                             --13
C: Clock(CLK, 10 ns, 10 ns);                                      --14
UUT: Mult8 port map (A, B, Start, CLK, Reset, Result, Done);      --15
DR <= Convert(Result);                                            --16
Reset  <= '1', '0' after 1 ns;                                    --17
process begin                                                     --18
   for i in 1 to 3 loop for j in 4 to 7 loop                      --19
      DA <= i; DB <= j;                                           --20
      A<=Convert(i,A'Length);B<=Convert(j,B'Length);              --21
      wait until CLK'EVENT and CLK='1'; wait for 1 ns;            --22
      Start <= '1', '0' after 20 ns; wait until Done = '1';       --23
      wait until CLK'EVENT and CLK='1';                           --24
   end loop; end loop;                                            --25
   for i in 0 to 1 loop for j in 0 to 15 loop                     --26
      DA <= i; DB <= j;                                           --27
      A<=Convert(i,A'Length);B<=Convert(j,B'Length);              --28
      wait until CLK'EVENT and CLK='1'; wait for 1 ns;            --29
      Start <= '1', '0' after 20 ns; wait until Done = '1';       --30
      wait until CLK'EVENT and CLK='1';                           --31
   end loop; end loop;                                            --32
   wait;                                                          --33
```

```
end process;                                                    --34
end;                                                            --35
```

Here is the signal trace output from the Compass Scout simulator:

```
        Time(fs) + Cycle            da          db          dr
--------------------- ------------ ------------ ------------
               0+ 0:           0           0           0
               0+ 1: *         1 *         4 *         0
...
        92000000+ 3:           1           4 *         4
...
       150000000+ 1: *         1 *         5           4
...
       193000000+ 3:           1           5 *         0
...
       252000000+ 3:           1           5 *         5
...
       310000000+ 1: *         1 *         6           5
...
       353000000+ 3:           1           6 *         0
...
       412000000+ 3:           1           6 *         6
```

Positive clock edges occur at 10, 30, 50, 70, 90, ... ns. You can see that the output (dr) changes from '0' to '4' at 92 ns, after five clock edges (with a 2 ns delay due to the output register, R1).

10.3   Syntax and Semantics of VHDL

We might define the **syntax** of a very small subset of the English language in **Backus–Naur form** (**BNF**) using **constructs** as follows:

```
sentence  ::= subject verb object.
subject   ::= The|A noun
object    ::= [article] noun {, and article noun}
article   ::= the|a
noun      ::= man|shark|house|food
verb      ::= eats|paints

::= means "can be replaced by"
|   means "or"
[]  means "contents optional"
{}  means "contents can be left out, used once, or repeated"
```

The following two English sentences are correct according to these syntax rules:

```
A shark eats food.
The house paints the shark, and the house, and a man.
```

We need **semantic rules** to tell us that the second sentence does not make much sense. Most of the VHDL LRM is dedicated to the definition of the language semantics. Appendix A of the LRM (which is not officially part of the standard) explains the complete VHDL syntax using BNF.

The rules that determine the characters you can use (the "alphabet" of VHDL), where you can put spaces, and so on are **lexical rules** [VHDL LRM13]. Any VHDL description may be written using a subset of the VHDL character set:

```
basic_character ::= upper_case_letter|digit|special_character
   |space_character|format_effector
```

The two space characters are: space (SP) and the nonbreaking space (NBSP). The five format effectors are: horizontal tabulation (HT), vertical tabulation (VT), carriage return (CR), line feed (LF), and form feed (FF). The characters that are legal in VHDL constructs are defined as the following subsets of the complete character set:

```
graphic_character ::=                                              [10.1]
    upper_case_letter|digit|special_character|space_character
   |lower_case_letter|other_special_character
```

```
special_character ::= " # & ' () * + , - . / : ; < = > [ ] _ |     [10.2]
```

The 11 other special characters are: ! $ % @ ? \ ^ ` { } ~, and (in VHDL-93 only) 34 other characters from the ISO Latin-1 set [ISO, 1987]. If you edit code using a word processor, you either need to turn smart quotes off or override this feature (use Tools... Preferences... General in MS Word; and use CTRL-' and CTRL-" in Frame).

When you learn a language it is difficult to understand how to use a noun without using it in a sentence. Strictly this means that we ought to define a sentence before we define a noun and so on. In this chapter I shall often break the "Define it before you use it" rule and use code examples and BNF definitions that contain VHDL constructs that we have not yet defined. This is often frustrating. You can use the book index and the table of important VHDL constructs at the end of this chapter (Table 10.28) to help find definitions if you need them.

We shall occasionally refer to the VHDL BNF syntax definitions in this chapter using references—BNF [10.1], for example. Only the most important BNF constructs for VHDL are included here in this chapter, but a complete description of the VHDL language syntax is contained in Appendix A.

10.4   Identifiers and Literals

Names (the "nouns" of VHDL) are known as **identifiers** [VHDL LRM13.3]. The correct "spelling" of an identifier is defined in BNF as follows:

```
identifier ::=                                                    [10.3]
    letter {[underline] letter_or_digit}
    |\graphic_character{graphic_character}\
```

In this book an underline in VHDL BNF marks items that are new or that have changed in VHDL-93 from VHDL-87. The following are examples of identifiers:

```
s -- A simple name.
S -- A simple name, the same as s. VHDL is not case sensitive.
a_name -- Imbedded underscores are OK.
-- Successive underscores are illegal in names: Ill__egal
-- Names can't start with underscore: _Illegal
-- Names can't end with underscore: Illegal_
Too_Good -- Names must start with a letter.
-- Names can't start with a number: 2_Bad
\74LS00\ -- Extended identifier to break rules (VHDL-93 only).
VHDL \vhdl\ \VHDL\ -- Three different names (VHDL-93 only).
s_array(0) -- A static indexed name (known at analysis time).
s_array(i) -- A non-static indexed name, if i is a variable.
```

You may not use a reserved word as a declared identifier, and it is wise not to use units, special characters, and function names: ns, ms, FF, read, write, and so on. You may attach qualifiers to names as follows [VHDL LRM6]:

```
CMOS.all -- A selected or expanded name, all units in library CMOS.
Data'LEFT(1) -- An attribute name, LEFT is the attribute designator.
Data(24 downto 1) -- A slice name, part of an array: Data(31 downto 0)
Data(1) -- An indexed name, one element of an array.
```

Comments follow two hyphens '--' and instruct the analyzer to ignore the rest of the line. There are no multiline comments in VHDL. Tabs improve readability, but it is best not to rely on a tab as a space in case the tabs are lost or deleted in conversion. You should thus write code that is still legal if all tabs are deleted.

There are various forms of **literals** (fixed-value items) in VHDL [VHDL LRM13.4–13.7]. The following code shows some examples:

```
entity Literals_1 is end;
architecture Behave of Literals_1 is
begin process
   variable I1 : integer; variable R1 : real;
   variable C1 : CHARACTER; variable S16 : STRING(1 to 16);
   variable BV4: BIT_VECTOR(0 to 3);
   variable BV12 : BIT_VECTOR(0 to 11);
   variable BV16 : BIT_VECTOR(0 to 15);
```

```
  begin
-- Abstract literals are decimal or based literals.
-- Decimal literals are integer or real literals.
-- Integer literal examples (each of these is the same):
    I1 := 120000; I1 := 12e4; I1 := 120_000;
-- Based literal examples (each of these is the same):
    I1 := 2#1111_1111#; I1 := 16#FF#;
-- Base must be an integer from 2 to 16:
    I1 := 16:FF:; -- you may use a : if you don't have #
-- Real literal examples (each of these is the same):
    R1 := 120000.0; R1 := 1.2e5; R1 := 12.0E4;
-- Character literal must be one of the 191 graphic characters.
-- 65 of the 256 ISO Latin-1 set are non-printing control characters
    C1 := 'A'; C1 := 'a'; -- different from each other
-- String literal examples:
    S16 := "  string" & " literal";    -- concatenate long strings
    S16 := """Hello,"" I said!";        -- doubled quotes
    S16 := %  string literal%;          -- can use % instead of "
    S16 := %Sale: 50%% off!!!%;         -- doubled %
-- Bit-string literal examples:
    BV4  := B"1100";   -- binary bit-string literal
    BV12 := O"7777";   -- octal  bit-string literal
    BV16 := X"FFFF";   -- hex     bit-string literal
wait; end process; -- the wait prevents an endless loop
end;
```

10.5 Entities and Architectures

The highest-level VHDL construct is the **design file** [VHDL LRM11.1]. A design file contains **design units** that contain one or more **library units**. Library units in turn contain: entity, configuration, and package declarations (**primary units**); and architecture and package bodies (**secondary units**).

```
design_file ::=                                              [10.4]
   {library_clause|use_clause} library_unit
   {{library_clause|use_clause} library_unit}

library_unit ::= primary_unit|secondary_unit

primary_unit ::=                                             [10.5]
   entity_declaration|configuration_declaration|package_declaration

secondary_unit ::= architecture_body|package_body           [10.6]
```

Using the written language analogy: a VHDL library unit is a "book," a VHDL design file is a "bookshelf," and a VHDL library is a collection of bookshelves. A

VHDL primary unit is a little like the chapter title and contents that appear on the first page of each chapter in this book and a VHDL secondary unit is like the chapter contents (though this is stretching our analogy a little far).

I shall describe the very important concepts of entities and architectures in this section and then cover libraries, packages, and package bodies. You define an entity, a black box, using an **entity declaration** [VHDL LRM1.1]. This is the BNF definition:

```
entity_declaration ::=                                    [10.7]
entity identifier is
    [generic (formal_generic_interface_list);]
    [port (formal_port_interface_list);]
    {entity_declarative_item}
  [begin
    {[label:] [postponed] assertion ;
    |[label:] [postponed] passive_procedure_call ;
    |passive_process_statement}]
end [entity] [entity_identifier] ;
```

The following is an example of an entity declaration for a black box with two inputs and an output:

```
entity Half_Adder is
  port (X, Y : in BIT := '0'; Sum, Cout : out BIT); -- formals
end;
```

Matching the parts of this code with the constructs in BNF [10.7] you can see that the identifier is Half_Adder and that (X, Y: in BIT := '0'; Sum, Cout: out BIT) corresponds to (port_interface_list) in the BNF. The ports X, Y, Sum, and Cout are **formal ports** or **formals**. This particular entity Half_Adder does not use any of the other optional constructs that are legal in an entity declaration.

The **architecture body** [VHDL LRM1.2] describes what an entity does, or the contents of the black box (it is architecture body and not architecture declaration).

```
architecture_body ::=                                     [10.8]
  architecture identifier of entity_name is
    {block_declarative_item}
      begin
      {concurrent_statement}
  end [architecture] [architecture_identifier] ;
```

For example, the following architecture body (I shall just call it an architecture from now on) describes the contents of the entity Half_Adder:

```
architecture Behave of Half_Adder is
  begin Sum <= X xor Y; Cout <= X and Y;
end Behave;
```

We use the same signal names (the formals: Sum, X, Y, and Cout) in the architecture as we use in the entity (we say the signals of the "parent" entity are **visible** inside the architecture "child"). An architecture can refer to other entity–architecture pairs—so we can nest black boxes. We shall often refer to an entity–architecture pair as entity(architecture). For example, the architecture Behave of the entity Half_Adder is Half_Adder(Behave).

Why would we want to describe the outside of a black box (an entity) separately from the description of its contents (its architecture)? Separating the two makes it easier to move between different architectures for an entity (there must be at least one). For example, one architecture may model an entity at a behavioral level, while another architecture may be a structural model.

A structural model that uses an entity in an architecture must declare that entity and its interface using a **component declaration** as follows [VHDL LRM4.5]:

```
component_declaration ::=                                      [10.9]
    component identifier [is]
      [generic (local_generic_interface_list);]
      [port (local_port_interface_list);]
    end component [component_identifier];
```

For example, the following architecture, Netlist, is a structural version of the behavioral architecture, Behave:

```
architecture Netlist of Half_Adder is
component MyXor port (A_Xor,B_Xor : in BIT; Z_Xor : out BIT);
end component; -- component with locals
component MyAnd port (A_And,B_And : in BIT; Z_And : out BIT);
end component; -- component with locals
begin
   Xor1: MyXor port map (X, Y, Sum);     -- instance with actuals
   And1 : MyAnd port map (X, Y, Cout);   -- instance with actuals
end;
```

Notice that:

- We declare the components: MyAnd, MyXor and their **local ports** (or **locals**): A_Xor, B_Xor, Z_Xor, A_And, B_And, Z_And.
- We instantiate the components with **instance names**: And1 and Xor1.
- We connect instances using **actual ports** (or **actuals**): X, Y, Sum, Cout.

Next we define the entities and architectures that we shall use for the components MyAnd and MyXor. You can think of an entity–architecture pair (and its formal ports) as a data-book specification for a logic cell; the component (and its local ports) corresponds to a software model for the logic cell; and an instance (and its actual ports) is the logic cell.

We do not need to write VHDL code for `MyAnd` and `MyXor`; the code is provided as a **technology library** (also called an **ASIC vendor library** because it is often sold or distributed by the ASIC company that will manufacture the chip—the ASIC vendor—and not the software company):

```
-- These definitions are part of a technology library:
entity AndGate is
    port (And_in_1, And_in_2 : in BIT; And_out : out BIT); -- formals
end;

architecture Simple of AndGate is
    begin And_out <= And_in_1 and And_in_2;
end;

entity XorGate is
    port (Xor_in_1, Xor_in_2 : in BIT; Xor_out : out BIT); -- formals
end;

architecture Simple of XorGate is
    begin Xor_out <= Xor_in_1 xor Xor_in_2;
end;
```

If we keep the description of a circuit's interface (the `entity`) separate from its contents (the `architecture`), we need a way to link or **bind** them together. A **configuration declaration** [VHDL LRM1.3] binds entities and architectures.

```
configuration_declaration ::=                                    [10.10]
    configuration identifier of entity_name is
        {use_clause|attribute_specification|group_declaration}
        block_configuration
    end [configuration] [configuration_identifier] ;
```

An entity–architecture pair is a **design entity**. The following configuration declaration defines which design entities we wish to use and associates the formal ports (from the entity declaration) with the local ports (from the component declaration):

```
configuration Simplest of Half_Adder is
use work.all;
    for Netlist
        for And1 : MyAnd use entity AndGate(Simple)
            port map -- association: formals => locals
                (And_in_1 => A_And, And_in_2 => B_And, And_out => Z_And);
        end for;
        for Xor1 : MyXor use entity XorGate(Simple)
            port map
                (Xor_in_1 => A_Xor, Xor_in_2 => B_Xor, Xor_out => Z_Xor);
        end for;
    end for;
end;
```

Figure 10.1 diagrams the use of entities, architectures, components, and configurations. This figure seems very complicated, but there are two reasons that VHDL works this way:

- Separating the entity, architecture, component, and configuration makes it easier to reuse code and change libraries. All we have to do is change names in the port maps and configuration declaration.

- We only have to alter and reanalyze the configuration declaration to change which architectures we use in a model—giving us a fast debug cycle.

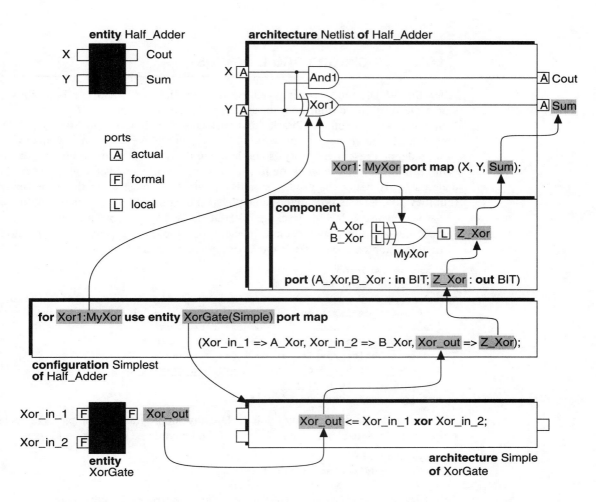

FIGURE 10.1   Entities, architectures, components, ports, port maps, and configurations.

You can think of design units, the analyzed entity–architecture pairs, as compiled object-code modules. The configuration then determines which object-code modules are linked together to form executable binary code.

You may also think of an entity as a block diagram, an architecture for an entity a more detailed circuit schematic for the block diagram, and a configuration as a parts list of the circuit components with their part numbers and manufacturers (also known as a **BOM** for **bill of materials,** rather like a shopping list). Most manufacturers (including the U.S. DoD) use schematics and BOMs as control documents for electronic systems. This is part of the rationale behind the structure of VHDL.

10.6 Packages and Libraries

After the VHDL tool has analyzed entities, architectures, and configurations, it stores the resulting design units in a library. Much of the power of VHDL comes from the use of predefined libraries and packages. A VHDL **design library** [VHDL LRM11.2] is either the current working library (things we are currently analyzing) or a predefined resource library (something we did yesterday, or we bought, or that came with the tool). The **working library** is named work and is the place where the code currently being analyzed is stored. Architectures must be in the same library (but they do not have to be in the same physical file on disk) as their parent entities.

You can use a VHDL **package** [VHDL LRM2.5–2.6] to define subprograms (procedures and functions), declare special types, modify the behavior of operators, or to hide complex code. Here is the BNF for a package declaration:

```
package_declaration ::=                                          [10.11]
package identifier is
{subprogram_declaration    | type_declaration    | subtype_declaration
   | constant_declaration    | signal_declaration    | file_declaration
   | alias_declaration       | component_declaration
   | attribute_declaration   | attribute_specification
   | disconnection_specification | use_clause
   | shared variable declaration | group_declaration
   | group_template_declaration}
end [package] [package_identifier] ;
```

You need a **package body** if you declare any subprograms in the package declaration (a package declaration and its body do not have to be in the same file):

```
package_body ::=
   package body package_identifier is
{subprogram_declaration    | subprogram_body
   | type_declaration       | subtype_declaration
   | constant_declaration   | file_declaration       | alias_declaration
   | use_clause
```

```
| shared_variable_declaration | group_declaration
| group_template_declaration}
end [package body] [package_identifier] ;
```

To make a package **visible** [VHDL LRM10.3] (or accessible, so you can see and use the package and its contents), you must include a **library clause** before a design unit and a **use clause** either before a design unit or inside a unit, like this:

```
library MyLib; -- library clause
use MyLib.MyPackage.all; -- use clause
-- design unit (entity + architecture, etc.) follows:
```

The STD and WORK libraries and the STANDARD package are always visible. Things that are visible to an entity are visible to its architecture bodies.

10.6.1 Standard Package

The VHDL **STANDARD package** [VHDL LRM14.2] is defined in the LRM and implicitly declares the following implementation dependent types: TIME, INTEGER, REAL. We shall use uppercase for types defined in an IEEE standard package. Here is part of the STANDARD package showing the explicit type and subtype declarations:

```
package Part_STANDARD is
type BOOLEAN is (FALSE, TRUE); type BIT is ('0', '1');
type SEVERITY_LEVEL is (NOTE, WARNING, ERROR, FAILURE);
subtype NATURAL is INTEGER range 0 to INTEGER'HIGH;
subtype POSITIVE is INTEGER range 1 to INTEGER'HIGH;
type BIT_VECTOR is array (NATURAL range <>) of BIT;
type STRING is array (POSITIVE range <>) of CHARACTER;
-- the following declarations are VHDL-93 only:
attribute FOREIGN: STRING; -- for links to other languages
subtype DELAY_LENGTH is TIME range 0 fs to TIME'HIGH;
type FILE_OPEN_KIND is (READ_MODE,WRITE_MODE,APPEND_MODE);
type FILE_OPEN_STATUS is
(OPEN_OK,STATUS_ERROR,NAME_ERROR,MODE_ERROR);
end Part_STANDARD;
```

Notice that a STRING array must have a positive index. The type TIME is declared in the STANDARD package as follows:

```
type TIME is range implementation_defined -- and varies with software
  units fs; ps = 1000 fs; ns = 1000 ps; us = 1000 ns; ms = 1000 us;
  sec = 1000 ms; min = 60 sec; hr = 60 min; end units;
```

The STANDARD package also declares the function now that returns the current simulation time (with type TIME in VHDL-87 and subtype DELAY_LENGTH in VHDL-93).

In VHDL-93 the CHARACTER type declaration extends the VHDL-87 declaration (the 128 ASCII characters):

```
type Part_CHARACTER is ( -- 128 ASCII characters in VHDL-87
NUL, SOH, STX, ETX, EOT, ENQ, ACK, BEL, -- 33 control characters
 BS,  HT,  LF,  VT,  FF,  CR,  SO,  SI, -- including:
DLE, DC1, DC2, DC3, DC4, NAK, SYN, ETB, -- format effectors:
CAN,  EM, SUB, ESC, FSP, GSP, RSP, USP, -- horizontal tab = HT
' ', '!', '"', '#', '$', '%', '&', ''', -- line feed = LF
'(', ')', '*', '+', ',', '-', '.', '/', -- vertical tab = VT
'0', '1', '2', '3', '4', '5', '6', '7', -- form feed = FF
'8', '9', ':', ';', '<', '=', '>', '?', -- carriage return = CR
'@', 'A', 'B', 'C', 'D', 'E', 'F', 'G', -- and others:
'H', 'I', 'J', 'K', 'L', 'M', 'N', 'O', -- FSP, GSP, RSP, USP use P
'P', 'Q', 'R', 'S', 'T', 'U', 'V', 'W', -- suffix to avoid conflict
'X', 'Y', 'Z', '[', '\', ']', '^', '_', -- with TIME units
'`', 'a', 'b', 'c', 'd', 'e', 'f', 'g',
'h', 'i', 'j', 'k', 'l', 'm', 'n', 'o',
'p', 'q', 'r', 's', 't', 'u', 'v', 'w',
'x', 'y', 'z', '{', '|', '}', '~', DEL  -- delete = DEL

-- VHDL-93 includes 96 more Latin-1 characters, like ¥ (Yen) and
-- 32 more control characters, better not to use any of them.
);
```

The VHDL-87 character set is the 7-bit coded **ISO 646-1983** standard known as the **ASCII character set**. Each of the printable ASCII graphic **character codes** (there are 33 nonprintable control codes, like DEL for delete) is represented by a **graphic symbol** (the shapes of letters on the keyboard, on the display, and that actually print). VHDL-93 uses the 8-bit coded character set **ISO 8859-1:1987(E)**, known as ISO **Latin-1**. The first 128 characters of the 256 characters in ISO Latin-1 correspond to the 128-character ASCII code. The graphic symbols for the printable ASCII characters are well defined, but not part of the standard (for example, the shape of the graphic symbol that represents 'lowercase a' is recognizable on every keyboard, display, and font). However, the graphic symbols that represent the printable characters from other 128-character codes of the ISO 8-bit character set are different in various fonts, languages, and computer systems. For example, a pound sterling sign in a U.K. character set looks like this–'£', but in some fonts the same character code prints as '#' (known as number sign, hash, or pound). If you use such characters and want to share your models with people in different countries, this can cause problems (you can see all 256 characters in a character set by using Insert... Symbol in MS Word).

10.6.2   Std_logic_1164 Package

VHDL does not have a built-in logic-value system. The STANDARD package predefines the type BIT with two logic values, '0' and '1', but we normally need at least two more values: `'X'` (unknown) and `'Z'` (high-impedance). Unknown is a **metalogical value** because it does not exist in real hardware but is needed for simulation purposes. We could define our own logic-value system with four logic values:

```
type MVL4 is ('X', '0', '1', 'Z'); -- a four-value logic system
```

The proliferation of VHDL logic-value systems prompted the creation of the **Std_logic_1164 package** (defined in IEEE Std 1164-1993) that includes functions to perform logical, shift, resolution, and conversion functions for types defined in the `Std_logic_1164` system. To use this package in a design unit, you must include the following library clause (before each design unit) and a use clause (either before or inside the unit):

```
library IEEE; use IEEE.std_logic_1164.all;
```

This `Std_Logic_1164` package contains definitions for a nine-value logic system. The following code and comments show the definitions and use of the most important parts of the package[2]:

```
package Part_STD_LOGIC_1164 is                                         --1
type STD_ULOGIC is                                                     --2
( 'U', -- Uninitialized                                                --3
  'X', -- Forcing Unknown                                              --4
  '0', -- Forcing 0                                                    --5
  '1', -- Forcing 1                                                    --6
  'Z', -- High Impedance                                               --7
  'W', -- Weak Unknown                                                 --8
  'L', -- Weak 0                                                       --9
  'H', -- Weak 1                                                       --10
  '-' -- Don't Care);                                                  --11
type STD_ULOGIC_VECTOR is array (NATURAL range <>) of STD_ULOGIC;      --12
function resolved (s : STD_ULOGIC_VECTOR) return STD_ULOGIC;           --13
subtype STD_LOGIC is resolved STD_ULOGIC;                              --14
type STD_LOGIC_VECTOR is array (NATURAL range <>) of STD_LOGIC;        --15
subtype X01  is resolved STD_ULOGIC range 'X' to '1';                  --16
subtype X01Z is resolved STD_ULOGIC range 'X' to 'Z';                  --17
subtype UX01 is resolved STD_ULOGIC range 'U' to '1';                  --18
subtype UX01Z is resolved STD_ULOGIC range 'U' to 'Z';                 --19

-- Vectorized overloaded logical operators:                            --20
function "and" (L : STD_ULOGIC; R : STD_ULOGIC) return UX01;           --21
-- Logical operators not, and, nand, or, nor, xor, xnor (VHDL-93),     --22
-- overloaded for STD_ULOGIC STD_ULOGIC_VECTOR STD_LOGIC_VECTOR.       --23

-- Strength strippers and type conversion functions:                  --24
-- function To_T (X : F) return T;                                     --25
-- defined for types, T and F, where                                  --26
-- F=BIT BIT_VECTOR STD_ULOGIC STD_ULOGIC_VECTOR STD_LOGIC_VECTOR      --27
-- T=types F plus types X01 X01Z UX01 (but not type UX01Z)             --28
```

---

[2]The code in this section is adapted with permission from IEEE Std 1164-1993, © Copyright IEEE. All rights reserved.

```
-- Exclude _'s in T in name: TO_STDULOGIC not TO_STD_ULOGIC        --29
-- To_XO1 : L->0, H->1 others->X                                   --30
-- To_XO1Z: Z->Z, others as To_X01                                 --31
-- To_UX01: U->U, others as To_X01                                 --32

-- Edge detection functions:                                       --33
function rising_edge (signal s: STD_ULOGIC) return BOOLEAN;        --34
function falling_edge (signal s: STD_ULOGIC) return BOOLEAN;       --35

-- Unknown detection (returns true if s = U, X, Z, W):             --36
-- function Is_X (s : T) return BOOLEAN;                           --37
-- defined for T = STD_ULOGIC STD_ULOGIC_VECTOR STD_LOGIC_VECTOR.  --38

end Part_STD_LOGIC_1164;                                           --39
```

Notice:

- The type STD_ULOGIC has nine logic values. For this reason IEEE Std 1164 is sometimes referred to as MVL9—multivalued logic nine. There are simpler, but nonstandard, MVL4 and MVL7 packages, as well as packages with more than nine logic values, available. Values 'U', 'X', and 'W' are all metalogical values.

- There are weak and forcing logic-value strengths. If more than one logic gate drives a node (there is more than one **driver**) as in wired-OR logic or a three-state bus, for example, the simulator checks the driver strengths to **resolve** the actual logic value of the node using the **resolution function**, resolved, defined in the package.

- The subtype STD_LOGIC is the **resolved** version of the **unresolved** type STD_ULOGIC. Since subtypes are **compatible** with types (you can assign one to the other) you can use either STD_LOGIC or STD_ULOGIC for a signal with a single driver, but it is generally safer to use STD_LOGIC.

- The type STD_LOGIC_VECTOR is the resolved version of unresolved type STD_ULOGIC_VECTOR. Since these are two different types and are not compatible, you should use STD_LOGIC_VECTOR. That way you will not run into a problem when you try to connect a STD_LOGIC_VECTOR to a STD_ULOGIC_VECTOR.

- The don't care logic value '-' (hyphen), is principally for use by synthesis tools. The value '-' is almost always treated the same as 'X'.

- The 1164 standard defines (or **overloads**) the logical operators for the STD_LOGIC types but not the arithmetic operators (see Section 10.12).

10.6.3   **TEXTIO** Package

You can use the **TEXTIO** package, which is part of the library STD, for text input and output [VHDL LRM14.3]. The following code is a part of the TEXTIO package header and, together with the comments, shows the declarations of types, subtypes, and the use of the procedures in the package:

```
package Part_TEXTIO is          -- VHDL-93 version.
type LINE is access STRING;     -- LINE is a pointer to a STRING value.
type TEXT is file of STRING;    -- File of ASCII records.
type SIDE is (RIGHT, LEFT);     -- for justifying output data.
subtype WIDTH is NATURAL;       -- for specifying widths of output fields.
file INPUT : TEXT open READ_MODE is "STD_INPUT"; -- Default input file.
file OUTPUT : TEXT open WRITE_MODE is "STD_OUTPUT"; -- Default output.

-- The following procedures are defined for types, T, where
-- T = BIT BIT_VECTOR BOOLEAN CHARACTER INTEGER REAL TIME STRING
--    procedure READLINE(file F : TEXT; L : out LINE);
--    procedure READ(L : inout LINE; VALUE : out T);
--    procedure READ(L : inout LINE; VALUE : out T; GOOD: out BOOLEAN);
--    procedure WRITELINE(F : out TEXT; L : inout LINE);
--    procedure WRITE(
--       L : inout LINE;
--       VALUE : in T;
--       JUSTIFIED : in SIDE:= RIGHT;
--       FIELD:in WIDTH := 0;
--       DIGITS:in NATURAL := 0;   -- for T = REAL only
--       UNIT:in TIME:= ns );      -- for T = TIME only
-- function ENDFILE(F : in TEXT) return BOOLEAN;

end Part_TEXTIO;
```

Here is an example that illustrates how to write to the screen (STD_OUTPUT):

```
library std; use std.textio.all; entity Text is end;
architecture Behave of Text is signal count : INTEGER := 0;
begin count <= 1 after 10 ns, 2 after 20 ns, 3 after 30 ns;
process (count) variable L: LINE; begin
if (count > 0) then
  write(L, now);                    -- Write time.
  write(L, STRING'(" count="));     -- STRING' is a type qualification.
  write(L, count); writeline(output, L);
end if; end process; end;

10 ns count=1
20 ns count=2
30 ns count=3
```

10.6.4   Other Packages

VHDL does not predefine arithmetic operators on types that hold bits. Many VHDL simulators provide one or more **arithmetic packages** that allow you to perform arithmetic operations on std_logic_1164 types. Some companies also provide one or more **math packages** that contain functions for floating-point algebra, trigonometry, complex algebra, queueing, and statistics (see also [IEEE 1076.2, 1996]).

Synthesis tool companies often provide a special version of an arithmetic package, a **synthesis package**, that allows you to synthesize VHDL that includes arithmetic operators. This type of package may contain special instructions (normally comments that are recognized by the synthesis software) that map common functions (adders, subtracters, multipliers, shift registers, counters, and so on) to ASIC library cells. I shall introduce the IEEE synthesis package in Section 10.12.

Synthesis companies may also provide **component packages** for such cells as power and ground pads, I/O buffers, clock drivers, three-state pads, and bus keepers. These components may be technology-independent (generic) and are mapped to primitives from technology-dependent libraries after synthesis.

10.6.5   Creating Packages

It is often useful to define constants in one central place rather than using literals wherever you need a specific value in your code. One way to do this is by using VHDL **packaged constants** [VHDL LRM4.3.1.1] that you define in a package. Packages that you define are initially part of the working library, work. Here are two example packages [VHDL LRM2.5–2.7]:

```
package Adder_Pkg is -- a package declaration
   constant BUSWIDTH : INTEGER := 16;
end Adder_Pkg;

use work.Adder_Pkg.all; -- a use clause
entity Adder is end Adder;
architecture Flexible of Adder is -- work.Adder_Pkg is visible here
   begin process begin
      MyLoop : for j in 0 to BUSWIDTH loop -- adder code goes here
      end loop; wait; -- the wait prevents an endless cycle
   end process;
end Flexible;

package GLOBALS is
   constant HI : BIT := '1'; constant LO: BIT := '0';
end GLOBALS;
```

Here is a package that declares a function and thus requires a package body:

```
package Add_Pkg_Fn is
function add(a, b, c : BIT_VECTOR(3 downto 0)) return BIT_VECTOR;
end Add_Pkg_Fn;

package body Add_Pkg_Fn is
function add(a, b, c : BIT_VECTOR(3 downto 0)) return BIT_VECTOR is
   begin return a xor b xor c; end;
end Add_Pkg_Fn;
```

The following example is similar to the **VITAL (VHDL Initiative Toward ASIC Libraries)** package that provides two alternative methods (procedures or functions) to model primitive gates (I shall describe functions and procedures in more detail in Section 10.9.2):

```
package And_Pkg is
   procedure V_And(a, b : BIT; signal c : out BIT);
   function V_And(a, b : BIT) return BIT;
end;

package body And_Pkg is
   procedure V_And(a, b : BIT; signal c : out BIT) is
      begin c <= a and b; end;
   function V_And(a, b : BIT) return BIT is
      begin return a and b; end;
end And_Pkg;
```

The software determines where it stores the design units that we analyze. Suppose the package Add_Pkg_Fn is in library MyLib. Then we need a library clause (before each design unit) and use clause with a selected name to use the package:

```
library MyLib; -- use MyLib.Add_Pkg.all; -- use all the package
use MyLib.Add_Pkg_Fn.add; -- just function 'add' from the package

entity Lib_1 is port (s : out BIT_VECTOR(3 downto 0) := "0000"); end;
architecture Behave of Lib_1 is begin process
begin s <= add ("0001", "0010", "1000"); wait; end process; end;
```

The VHDL software dictates how you create the library MyLib from the library work and the actual name and directory location for the physical file or directory on the disk that holds the library. The mechanism to create the links between the file and directory names in the computer world and the library names in the VHDL world depends on the software. There are three common methods:

- Use a UNIX environment variable (SETENV MyLib ~/MyDirectory/ MyLibFile, for example).

- Create a separate file that establishes the links between the filename known to the operating system and the library name known to the VHDL software.

- Include the links in an initialization file (often with an '.ini' suffix).

10.7 Interface Declarations

An **interface declaration** declares **interface objects** that may be interface constants, signals, variables, or files [VHDL 87LRM4.3.3, 93LRM4.3.2]. **Interface constants** are generics of a design entity, a component, or a block, or parameters of subprograms. **Interface signals** are ports of a design entity, component, or block, and parameters of subprograms. **Interface variables** and **interface files** are parameters of subprograms.

Each interface object has a **mode** that indicates the direction of information flow. The most common modes are in (the default), out, inout, and buffer (a fifth mode, linkage, is used to communicate with other languages and is infrequently used in ASIC design). The restrictions on the use of objects with these modes are listed in Table 10.10. An interface object is **read** when you use it on the RHS of an assignment statement, for example, or when the object is associated with another interface object of modes in, inout (or linkage). An interface object is **updated** when you use it on the LHS side of an assignment statement or when the object is associated with another interface object of mode out, buffer, inout (or linkage). The restrictions on reading and updating objects generate the diagram at the bottom of Table 10.10 that shows the 10 allowed types of interconnections (these rules for modes buffer and inout are the same). The interface objects (Inside and Outside) in the example in this table are ports (and thus interface signals), but remember that interface objects may also be interface constants, variables, and files.

There are other special-case rules for reading and updating interface signals, constants, variables, and files that I shall cover in the following sections. The situation is like the spelling rule, "i before e except after c." Table 10.10 corresponds to the rule "i before e."

10.7.1   Port Declaration

Interface objects that are signals are called **ports** [VHDL 93LRM1.1.1.2]. You may think of ports as "connectors" and you must declare them as follows:

port (*port*_interface_list)

```
interface_list ::=                                              [10.12]
   port_interface_declaration {; port_interface_declaration}
```

A **port interface declaration** is a list of ports that are the inputs and outputs of an entity, a block, or a component declaration:

```
interface_declaration ::=                                       [10.13]
   [signal]
     identifier {, identifier}:[in|out|inout|buffer|linkage]
     subtype_indication [bus] [:= static_expression]
```

Each port forms an **implicit signal declaration** and has a **port mode**. I shall discuss bus, which is a **signal kind**, in Section 10.13.1. Here is an example of an entity declaration that has five ports:

```
entity Association_1 is
   port (signal X, Y : in BIT := '0'; Z1, Z2, Z3 : out BIT);
end;
```

TABLE 10.10   Modes of interface objects and their properties.

```
entity E1 is port (Inside : in BIT); end; architecture Behave of E1 is begin end;
entity E2 is port (Outside : inout BIT := '1'); end; architecture Behave of E2 is
component E1 port (Inside: in BIT); end component; signal UpdateMe : BIT; begin
I1 : E1 port map (Inside => Outside);   -- formal/local (mode in) => actual (mode inout)
UpdateMe <= Outside;                    -- OK to read Outside (mode inout)
Outside  <= '0' after 10 ns;            -- and OK to update Outside (mode inout)
end;
```

Possible modes of interface object, Outside	in (default)	out	inout	buffer
Can you read Outside (RHS of assignment)?	Yes	No	Yes	Yes
Can you update Outside (LHS of assignment)?	No	Yes	Yes	Yes
Modes of Inside that Outside may connect to (see below)[1]	in	out	any	any

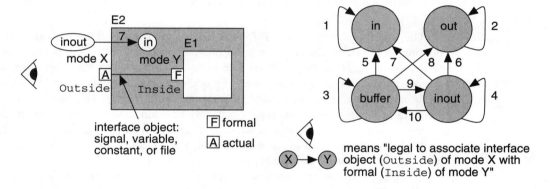

F formal
A actual

interface object:
signal, variable,
constant, or file

means "legal to associate interface
object (Outside) of mode X with
formal (Inside) of mode Y"

[1]There are additional rules for interface objects that are signals (ports)—see Tables 10.11 and 10.12.

In the preceding declaration the keyword signal is redundant (because all ports are signals) and may be omitted. You may also omit the port mode in because it is the default mode. In this example, the input ports X and Y are driven by a **default value** (in general a **default expression**) of '0' if (and only if) the ports are left unconnected or **open**. If you do leave an input port open, the port must have a default expression.

You use a **port map** and either **positional association** or **named association** to connect the formals of an entity with the locals of a component. Port maps also associate (connect) the locals of a component with the actuals of an instance. For an example of formal, local, and actual ports, and explanation of their function, see Section 10.5, where we declared an entity AndGate. The following example shows how to bind a component to the entity AndGate (in this case we use the **default binding**) and associate the ports. Notice that if we mix positional and named association then all positional associations must come first.

```
use work.all; -- makes analyzed design entity AndGate(Simple) visible.
architecture Netlist of Association_1 is
-- The formal port clause for entity AndGate looks like this:
-- port (And_in_1, And_in_2: in BIT; And_out : out BIT); -- Formals.
component AndGate port
    (And_in_1, And_in_2 : in BIT; And_out : out BIT); -- Locals.
end component;
begin
-- The component and entity have the same names: AndGate.
-- The port names are also the same: And_in_1, And_in_2, And_out,
-- so we can use default binding without a configuration.
-- The last (and only) architecture for AndGate will be used: Simple.
A1:AndGate port map (X, Y, Z1); -- positional association
A2:AndGate port map (And_in_2=>Y, And_out=>Z2, And_in_1=>X);   -- named
A3:AndGate port map (X, And_out => Z3, And_in_2 => Y);         -- both
end;
```

The interface object rules of Table 10.10 apply to ports. The rule that forbids updating an interface object of mode in prevents modifying an input port (by placing the input signal on the left-hand side of an assignment statement, for example). Less obviously, you cannot read a port of mode out (that is you cannot place an output signal on the right-hand side of an assignment statement). This stops you from accidentally reading an output signal that may be connected to a net with multiple drivers. In this case the value you would read (the unresolved output signal) might not be the same as the resolved signal value. For example, in the following code, since Clock is a port of mode out, you cannot read Clock directly. Instead you can transfer Clock to an intermediate variable and read the intermediate variable instead:

```
entity ClockGen_1 is port (Clock : out BIT); end;
architecture Behave of ClockGen_1 is
begin process variable Temp : BIT := '1';
  begin
-- Clock <= not Clock;   -- Illegal, you cannot read Clock (mode out),
  Temp := not Temp;      -- use a temporary variable instead.
  Clock <= Temp after 10 ns; wait for 10 ns;
  if (now > 100 ns) then wait; end if; end process;
end;
```

Table 10.10 lists the restrictions on reading and updating interface objects including interface signals that form ports. Table 10.11 lists additional special rules for reading and updating the attributes of interface signals.

TABLE 10.11   Properties of ports.

Example entity declaration:
entity E **is port** (F_1:BIT; F_2:**out** BIT; F_3:**inout** BIT; F_4:**buffer** BIT); **end;**   -- formals

Example component declaration:
component C **port** (L_1:BIT; L_2:**out** BIT; L_3:**inout** BIT; L_4:**buffer** BIT); -- locals
end component;

Example component instantiation:
I1 : C **port map**
(L_1 => A_1, L_2 => A_2, L_3 => A_3, L_4 => A_4); -- locals => actuals

Example configuration:
for I1 : C **use entity** E(Behave) **port map**
(F_1 => L_1, F_2 => L_2, F_3 => L_3, F_4 => L_4); -- formals => locals

Interface object, port F	F_1	F_2	F_3	F_4
Mode of F	**in** (default)	**out**	**inout**	**buffer**
Can you read attributes of F? [VHDL LRM4.3.2]	Yes, but not the attributes: 'STABLE 'QUIET 'DELAYED 'TRANSACTION	Yes, but not the attributes: 'STABLE 'QUIET 'DELAYED 'TRANSACTION 'EVENT 'ACTIVE 'LAST_EVENT 'LAST_ACTIVE 'LAST_VALUE	Yes, but not the attributes: 'STABLE 'QUIET 'DELAYED 'TRANSACTION	Yes

There is one more set of rules that apply to port connections [VHDL LRM 1.1.1.2]. If design entity E2 contains an instance, I1, of design entity E1, then the formals (of design entity E1) are associated with actuals (of instance I1). The actuals (of instance I1) are themselves formal ports (of design entity E2). The restrictions illustrated in Table 10.12 apply to the modes of the port connections from E1 to E2 (looking from the inside to the outside).

Notice that the allowed connections diagrammed in Table 10.12 (looking from inside to the outside) are a superset of those of Table 10.10 (looking from the outside to the inside). Only the seven types of connections shown in Table 10.12 are allowed between the ports of nested design entities. The additional rule that ports of mode buffer may only have one source, together with the restrictions on port mode interconnections, limits the use of ports of mode buffer.

TABLE 10.12   Connection rules for port modes.

```
entity E1 is port (Inside : in BIT); end; architecture Behave of E1 is begin end;
entity E2 is port (Outside : inout BIT := '1'); end; architecture Behave of E2 is
component E1 port (Inside : in BIT); end component; begin
I1 : E1 port map (Inside => Outside);   -- formal/local (mode in) => actual (mode inout)
end;
```

Possible modes of interface object, Inside	**in** (default)	**out**	inout	buffer
Modes of Outside that Inside may connect to (see below)	**in inout buffer**	**out inout**	**inout**[1]	**buffer**[2]

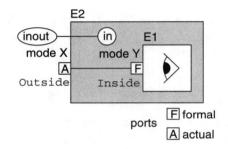

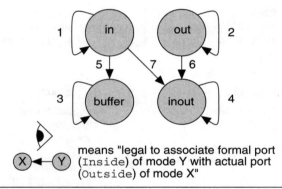

ports $\boxed{\text{F}}$ formal
  $\boxed{\text{A}}$ actual

means "legal to associate formal port (Inside) of mode Y with actual port (Outside) of mode X"

[1]A signal of mode inout can be updated by any number of sources [VHDL 87LRM 4.3.3, 93LRM4.3.2].
[2]A signal of mode buffer can be updated by at most one source [VHDL LRM1.1.1.2].

10.7.2   Generics

Ports are signals that carry changing information between entities. A **generic** is similar to a port, except generics carry constant, static information. A generic is an interface constant that, unlike normal VHDL constants, may be given a value in a component instantiation statement or in a configuration specification. You declare generics in an entity declaration and you use generics in a similar fashion to ports. The following example uses a generic parameter to alter the size of a gate:

```
entity AndGateNWide is
   generic (N : NATURAL := 2);
   port (Inputs : BIT_VECTOR(1 to N); Result : out BIT);
end;
```

Notice that the **generic interface list** precedes the port interface list. Generics are useful to carry timing (delay) information, as in the next example:

```
entity AndT is
   generic (TPD : TIME := 1 ns);
   port (a, b : BIT := '0'; q: out BIT);
```

```
end;
architecture Behave of AndT is
  begin q <= a and b after TPD;
end;

entity AndT_Test_1 is end;
architecture Netlist_1 of AndT_Test_1 is
  component MyAnd
    port (a, b : BIT; q : out BIT);
  end component;
  signal a1, b1, q1 : BIT := '1';
  begin
    And1 : MyAnd port map (a1, b1, q1);
end Netlist_1;

configuration Simplest_1 of AndT_Test_1 is use work.all;
  for Netlist_1 for And1 : MyAnd
    use entity AndT(Behave) generic map (2 ns);
  end for; end for;
end Simplest_1;
```

The configuration declaration, Simplest_1, changes the default delay (equal to 1 ns, declared as a default expression in the entity) to 2 ns. Techniques based on this method are useful in ASIC design. Prelayout simulation uses the default timing values. Back-annotation alters the delay in the configuration for postlayout simulation. When we change the delay we only need to reanalyze the configuration, not the rest of the ASIC model.

There was initially no standard in VHDL for how timing generics should be used, and the lack of a standard was a major problem for ASIC designers. The IEEE 1076.4 VITAL standard addresses this problem (see Section 13.5.5).

10.8 Type Declarations

In some programming languages you must declare objects to be integer, real, Boolean, and so on. VHDL (and ADA, the DoD programming language to which VHDL is related) goes further: You must declare the **type** of an object, and there are strict rules on mixing objects of different types. We say VHDL is strongly typed. For example, you can use one type for temperatures in Centigrade and a different type for Fahrenheit, even though both types are real numbers. If you try to add a temperature in Centigrade to a temperature in Fahrenheit, VHDL catches your error and tells you that you have a type mismatch.

This is the formal (expanded) BNF definition of a **type declaration**:

```
type_declaration ::=                                    [10.14]
  type identifier ;
| type identifier is
```

```
(identifier|'graphic_character' {, identifier|'graphic_character'}) ;
| range_constraint ;        | physical_type_definition ;
| record_type_definition ;  | access subtype_indication ;
| file of type_name ;       | file of subtype_name ;
| array index_constraint of element_subtype_indication ;
| array
  (type_name|subtype_name range <>
    {, type_name|subtype_name range <>}) of
    element_subtype_indication ;
```

There are four **type classes** in VHDL [VHDL LRM3]: **scalar types, composite types, access types**, and **file types**. The scalar types are: **integer type, floating-point type, physical type**, and **enumeration type**. Integer and enumeration types are **discrete types**. Integer, floating-point, and physical types are **numeric types**. The **range** of an integer is implementation dependent but is guaranteed to include -2147483647 to $+2147483647$. Notice the integer range is symmetric and equal to $-(2^{31}-1)$ to $(2^{31}-1)$. Floating-point size is implementation dependent, but the range includes the bounds $-1.0E38$ and $+1.0E38$, and must include a minimum of six decimal digits of precision. Physical types correspond to time, voltage, current, and so on and have dimensions—a unit of measure (seconds, for example). Access types are pointers, useful in abstract data structures, but less so in ASIC design. File types are used for file I/O.

You may also declare a subset of an existing type, known as a **subtype**, in a **subtype declaration**. We shall discuss the different treatment of types and subtypes in expressions in Section 10.12.

Here are some examples of scalar type [VHDL LRM4.1] and subtype declarations [VHDL LRM4.2]:

```
entity Declaration_1 is end; architecture Behave of Declaration_1 is
type F is range 32 to 212; -- Integer type, ascending range.
type C is range 0 to 100; -- Range 0 to 100 is the range constraint.
subtype G is INTEGER range 9 to 0; -- Base type INTEGER, descending.
-- This is illegal: type Bad100 is INTEGER range 0 to 100;
-- don't use INTEGER in declaration of type (but OK in subtype).
type Rainbow is (R, O, Y, G, B, I, V); -- An enumeration type.
-- Enumeration types always have an ascending range.
type MVL4 is ('X', '0', '1', 'Z');
-- Note that 'X' and 'x' are different character literals.
-- The default initial value is MVL4'LEFT = 'X'.
-- We say '0' and '1' (already enumeration literals
-- for predefined type BIT) are overloaded.
-- Illegal enumeration type: type Bad4 is ("X", "0", "1", "Z");
-- Enumeration literals must be character literals or identifiers.
begin end;
```

The most common composite type is the **array type** [VHDL LRM3.2.1]. The following examples illustrate the semantics of array declarations:

```
entity Arrays_1 is end; architecture Behave of Arrays_1 is
type Word is array (0 to 31) of BIT; -- a 32-bit array, ascending
type Byte is array (NATURAL range 7 downto 0) of BIT; -- descending
type BigBit is array (NATURAL range <>) of BIT;
-- We call <> a box, it means the range is undefined for now.
-- We call BigBit an unconstrained array.
-- This is OK, we constrain the range of an object that uses
-- type BigBit when we declare the object, like this:
subtype Nibble is BigBit(3 downto 0);
type T1 is array (POSITIVE range 1 to 32) of BIT;
-- T1, a constrained array declaration, is equivalent to a type T2
-- with the following three declarations:
subtype index_subtype is POSITIVE range 1 to 32;
type array_type is array (index_subtype range <>) of BIT;
subtype T2 is array_type (index_subtype);
-- We refer to index_subtype and array_type as being
-- anonymous subtypes of T1 (since they don't really exist).
begin end;
```

You can assign values to an array using **aggregate notation** [VHDL LRM7.3.2]:

```
entity Aggregate_1 is end; architecture Behave of Aggregate_1 is
type D is array (0 to 3) of BIT; type Mask is array (1 to 2) of BIT;
signal MyData : D := ('0', others => '1'); -- positional aggregate
signal MyMask : Mask := (2 => '0', 1 => '1'); -- named aggregate
begin end;
```

The other composite type is the **record type** that groups elements together:

```
entity Record_2 is end; architecture Behave of Record_2 is
type Complex is record real : INTEGER; imag : INTEGER; end record;
signal s1 : Complex := (0, others => 1); signal s2: Complex;
begin s2 <= (imag => 2, real => 1); end;
```

10.9   Other Declarations

A declaration is one of the following [VHDL LRM4]:

```
declaration ::=                                              [10.15]
  type_declaration      | subtype_declaration | object_declaration
| interface_declaration | alias_declaration   | attribute_declaration
| component_declaration | entity_declaration
| configuration_declaration
| subprogram_declaration | package_declaration
| group_template_declaration | group_declaration
```

I discussed entity, configuration, component, package, interface, type, and sub-type declarations in Sections 10.5–10.8. Next I shall discuss the other types of declarations (except for groups or group templates [VHDL 93LRM4.6–4.7], new to VHDL-93, that are not often used in ASIC design).

10.9.1   Object Declarations

There are four **object classes** in VHDL: **constant**, **variable**, **signal**, and **file** [VHDL LRM 4.3.1.1-4.3.1.3]. You use a **constant declaration**, **signal declaration**, **variable declaration**, or **file declaration** together with a type. Signals can only be declared in the **declarative region** (before the first `begin`) of an architecture or block, or in a package (not in a package body). Variables can only be declared in the declarative region of a process or subprogram (before the first `begin`). You can think of signals as representing real wires in hardware. You can think of variables as memory locations in the computer. Variables are more efficient than signals because they require less overhead.

You may assign an (explicit) **initial value** when you declare a type. If you do not provide initial values, the (implicit) **default initial value** of a type or subtype `T` is `T'LEFT` (the leftmost item in the range of the type). For example:

```
entity Initial_1 is end; architecture Behave of Initial_1 is
type Fahrenheit is range 32 to 212;    -- Default initial value is 32.
type Rainbow is (R, O, Y, G, B, I, V); -- Default initial value is R.
type MVL4 is ('X', '0', '1', 'Z');     -- MVL4'LEFT = 'X'.
begin end;
```

The details of initialization and assignment of initial values are important—it is difficult to implement the assignment of initial values in hardware—instead it is better to mimic the hardware and use explicit reset signals.

Here are the formal definitions of constant and signal declarations:

```
constant_declaration ::= constant                                        [10.16]
identifier {, identifier}:subtype_indication [:= expression] ;
```

```
signal_declaration ::= signal                                            [10.17]
identifier {, identifier}:subtype_indication [register|bus] [:=expression];
```

I shall explain the use of signals of kind `register` or bus in Section 10.13.1. Signal declarations are **explicit signal declarations** (ports declared in an interface declaration are implicit signal declarations). Here is an example that uses a constant and several signal declarations:

```
entity Constant_2 is end;
library IEEE; use IEEE.STD_LOGIC_1164.all;
architecture Behave of Constant_2 is
constant Pi : REAL := 3.14159;              -- A constant declaration.
signal B : BOOLEAN; signal s1, s2: BIT;
signal sum : INTEGER range 0 to 15;         -- Not a new type.
signal SmallBus : BIT_VECTOR (15 downto 0); -- 16-bit bus.
```

```
signal GBus : STD_LOGIC_VECTOR (31 downto 0) bus; -- A guarded signal.
begin end;
```

Here is the formal definition of a variable declaration:

```
variable_declaration ::= [shared] variable                         [10.18]
identifier {, identifier}:subtype_indication [:= expression] ;
```

A **shared variable** can be used to model a varying quantity that is common across several parts of a model, temperature, for example, but shared variables are rarely used in ASIC design. The following examples show that variable declarations belong inside a process statement, after the keyword process and before the first appearance of the keyword begin inside a process:

```
library IEEE; use IEEE.STD_LOGIC_1164.all; entity Variables_1 is end;
architecture Behave of Variables_1 is begin process
   variable i : INTEGER range 1 to 10 := 10; -- Initial value = 10.
   variable v : STD_LOGIC_VECTOR (0 to 31) := (others => '0');
   begin wait; end process; -- The wait stops an endless cycle.
end;
```

10.9.2 Subprogram Declarations

VHDL code that you use several times can be declared and specified as **subprograms** (functions or procedures) [VHDL LRM2.1]. A **function** is a form of expression, may only use parameters of mode in, and may not contain delays or sequence events during simulation (no wait statements, for example). Functions are useful to model combinational logic. A **procedure** is a form of statement and allows you to control the scheduling of simulation events without incurring the overhead of defining several separate design entities. There are thus two forms of **subprogram declaration**: a **function declaration** or a **procedure declaration**.

```
subprogram_declaration ::= subprogram_specification ; ::=         [10.19]
 procedure
   identifier|string_literal [(parameter_interface_list)]
| [pure|impure] function
   identifier|string_literal [(parameter_interface_list)]
return type_name|subtype_name;
```

Here are a function and a procedure declaration that illustrate the difference:

```
function add(a, b, c : BIT_VECTOR(3 downto 0)) return BIT_VECTOR is
-- A function declaration, a function can't modify a, b, or c.

procedure Is_A_Eq_B (signal A, B : BIT; signal Y : out BIT);
-- A procedure declaration, a procedure can change Y.
```

Parameter names in subprogram declarations are called **formal parameters** (or formals). During a call to a subprogram, known as **subprogram invocation**, the passed values are **actual parameters** (or actuals). An **impure** function, such as the function now or a function that writes to or reads from a file, may return different values each time it is called (even with the same actuals). A **pure** function (the default) returns the same value if it is given the same actuals. You may call subprograms recursively. Table 10.13 shows the properties of subprogram parameters.

TABLE 10.13   Properties of subprogram parameters.

Example subprogram declarations:
```
function my_function(Ff) return BIT is -- Formal function parameter, Ff.
procedure my_procedure(Fp);         -- Formal procedure parameter, Fp.
```

Example subprogram calls:
```
my_result := my_function(Af); -- Calling a function with an actual parameter, Af.
MY_LABEL:my_procedure(Ap);    -- Using a procedure with an actual parameter, Ap.
```

Mode of `Ff` or `Fp` (formals)	`in`	`out`	`inout`	No mode
Permissible classes for `Af` (function actual parameter)	`constant` (default) `signal`	Not allowed	Not allowed	`file`
Permissible classes for `Ap` (procedure actual parameter)	`constant` (default) `variable` `signal`	`constant` `variable` (default) `signal`	`constant` `variable` (default) `signal`	`file`
Can you read attributes of `Ff` or `Fp` (formals)?	Yes, except: `'STABLE` `'QUIET` `'DELAYED` `'TRANSACTION` of a signal	Yes, except: `'STABLE` `'QUIET` `'DELAYED` `'TRANSACTION` `'EVENT` `'ACTIVE` `'LAST_EVENT` `'LAST_ACTIVE` `'LAST_VALUE` of a signal	Yes, except: `'STABLE` `'QUIET` `'DELAYED` `'TRANSACTION` of a signal	

A subprogram declaration is optional, but a **subprogram specification** must be included in the **subprogram body** (and must be identical in syntax to the subprogram declaration—see BNF [10.19]):

```
subprogram_body ::=                                              [10.20]
    subprogram_specification is
    {subprogram_declaration|subprogram_body
    |type_declaration|subtype_declaration
    |constant_declaration|variable_declaration|file_declaration
```

```
 |alias_declaration|attribute_declaration|attribute_specification
 |use_clause|group_template_declaration|group_declaration}
begin
   {sequential_statement}
 end [procedure|function] [identifier|string_literal] ;
```

You can include a subprogram declaration or subprogram body in a package or package body (see Section 10.6) or in the declarative region of an entity or process statement. The following is an example of a function declaration and its body:

```
function subset0(sout0 : in BIT) return BIT_VECTOR -- declaration

-- Declaration can be separate from the body.

function subset0(sout0 : in BIT) return BIT_VECTOR is -- body
variable y : BIT_VECTOR(2 downto 0);
begin
if (sout0 = '0') then y := "000"; else y := "100"; end if;
return result;
end;

procedure clockGen (clk : out BIT)                  -- Declaration

procedure clockGen (clk : out BIT) is               -- Specification
begin -- Careful this process runs forever:
   process begin wait for 10 ns; clk <= not clk; end process;
end;
```

One reason for having the optional (and seemingly redundant) subprogram declaration is to allow companies to show the subprogram declarations (to document the interface) in a package declaration, but to hide the subprogram bodies (the actual code) in the package body. If a separate subprogram declaration is present, it must **conform** to the specification in the subprogram body [VHDL 93LRM2.7]. This means the specification and declaration must be almost identical; the safest method is to copy and paste. If you define common procedures and functions in packages (instead of in each entity or architecture, for example), it will be easier to reuse subprograms. In order to make a subprogram included in a package body visible outside the package, you must declare the subprogram in the package declaration (otherwise the subprogram is **private**).

You may call a function from any expression, as follows:

```
entity F_1 is port (s : out BIT_VECTOR(3 downto 0) := "0000"); end;
architecture Behave of F_1 is begin process
function add(a, b, c : BIT_VECTOR(3 downto 0)) return BIT_VECTOR is
begin return a xor b xor c; end;
begin s <= add("0001", "0010", "1000"); wait; end process; end;

package And_Pkg is
   procedure V_And(a, b : BIT; signal c : out BIT);
   function V_And(a, b : BIT) return BIT;
```

```
end;
package body And_Pkg is
  procedure V_And(a,b : BIT;signal c : out BIT) is
    begin c <= a and b; end;
  function V_And(a,b : BIT) return BIT is
    begin return a and b; end;
end And_Pkg;

entity F_2 is port (s: out BIT := '0'); end;
use work.And_Pkg.all; -- use package already analyzed
architecture Behave of F_2 is begin process begin
s <= V_And('1', '1'); wait; end process; end;
```

I shall discuss the two different ways to call a procedure in Sections 10.10.4 and 10.13.3.

10.9.3   Alias and Attribute Declarations

An **alias declaration** [VHDL 87LRM4.3.4, 93LRM4.3.3] names parts of a type:

```
alias_declaration ::=                                        [10.21]
alias
    identifier|character_literal|operator_symbol [ :subtype_indication]
    is name [signature];
```

(the subtype indication is required in VHDL-87, but not in VHDL-93).

Here is an example of alias declarations for parts of a floating-point number:

```
entity Alias_1 is end; architecture Behave of Alias_1 is
begin process variable Nmbr: BIT_VECTOR (31 downto 0);
-- alias declarations to split Nmbr into 3 pieces :
alias Sign :        BIT is Nmbr(31);
alias Mantissa :    BIT_VECTOR (23 downto 0) is Nmbr (30 downto 7);
alias Exponent :    BIT_VECTOR ( 6 downto 0) is Nmbr ( 6 downto 0);
begin wait; end process; end; -- the wait prevents an endless cycle
```

An **attribute declaration** [VHDL LRM4.4] defines attribute properties:

```
attribute_declaration ::=                                    [10.22]
 attribute identifier:type_name ; | attribute identifier:subtype_name ;
```

Here is an example:

```
entity Attribute_1 is end; architecture Behave of Attribute_1 is
begin process type COORD is record X, Y : INTEGER; end record;
attribute LOCATION : COORD; -- the attribute declaration
begin wait ; -- the wait prevents an endless cycle
end process; end;
```

You define the attribute properties in an **attribute specification** (the following example specifies an attribute of a component label). You probably will not need to use your own attributes very much in ASIC design.

```
attribute LOCATION of adder1 : label is (10,15);
```

You can then refer to your attribute as follows:

```
positionOfComponent := adder1'LOCATION;
```

10.9.4 Predefined Attributes

The predefined attributes for scalar and array types in VHDL-93 are shown in Table 10.14 [VHDL 93LRM14.1]. There are two attributes, 'STRUCTURE and 'BEHAVIOR, that are present in VHDL-87, but removed in VHDL-93. Both of these attributes apply to architecture bodies. The attribute name A'BEHAVIOR is TRUE if the architecture A does not contain component instantiations. The attribute name A'STRUCTURE is TRUE if the architecture A contains only passive processes (those with no assignments to signals) and component instantiations. These two attributes were not widely used. The attributes shown in Table 10.14, however, are used extensively to create packages and functions for type conversion and overloading operators, but should not be needed by an ASIC designer. Many of the attributes do not correspond to "real" hardware and cannot be implemented by a synthesis tool.

The attribute 'LEFT is important because it determines the default initial value of a type. For example, the default initial value for type BIT is BIT'LEFT, which is '0'. The predefined attributes of signals are listed in Table 10.15. The most important signal attribute is 'EVENT, which is frequently used to detect a clock edge. Notice that Clock'EVENT, for example, is a function that returns a value of type BOOLEAN, whereas the otherwise equivalent not(Clock'STABLE), is a signal. The difference is subtle but important when these attributes are used in the wait statement that treats signals and values differently.

10.10 Sequential Statements

A **sequential statement** [VHDL LRM8] is defined as follows:

```
sequential_statement ::=                                  [10.23]
   wait_statement     | assertion_statement
 | signal_assignment_statement
 | variable_assignment_statement     | procedure_call_statement
 | if_statement     | case_statement | loop_statement
 | next_statement     | exit_statement
 | return_statement     | null_statement | report_statement
```

TABLE 10.14 Predefined attributes for scalar and array types.

Attribute	Kind[1]	Prefix T, A, E[2]	Parameter X or N[3]	Result type[3]	Result
T'BASE	T	any		base(T)	base(T), use only with other attribute
T'LEFT	V	scalar		T	Left bound of T
T'RIGHT	V	scalar		T	Right bound of T
T'HIGH	V	scalar		T	Upper bound of T
T'LOW	V	scalar		T	Lower bound of T
T'ASCENDING	V	scalar		BOOLEAN	True if range of T is ascending[4]
T'IMAGE(X)	F	scalar	base(T)	STRING	String representation of X in T[4]
T'VALUE(X)	F	scalar	STRING	base(T)	Value in T with representation X[4]
T'POS(X)	F	discrete	base(T)	UI	Position number of X in T (starts at 0)
T'VAL(X)	F	discrete	UI	base(T)	Value of position X in T
T'SUCC(X)	F	discrete	base(T)	base(T)	Value of position X in T plus one
T'PRED(X)	F	discrete	base(T)	base(T)	Value of position X in T minus one
T'LEFTOF(X)	F	discrete	base(T)	base(T)	Value to the left of X in T
T'RIGHTOF(X)	F	discrete	base(T)	base(T)	Value to the right of X in T
A'LEFT[(N)]	F	array	UI	T(Result)	Left bound of index N of array A
A'RIGHT[(N)]	F	array	UI	T(Result)	Right bound of index N of array A
A'HIGH[(N)]	F	array	UI	T(Result)	Upper bound of index N of array A
A'LOW[(N)]	F	array	UI	T(Result)	Lower bound of index N of array A
A'RANGE[(N)]	R	array	UI	T(Result)	Range A'LEFT(N) to A'RIGHT(N)[5]
A'REVERSE_RANGE[(N)]	R	array	UI	T(Result)	Opposite range to A'RANGE[(N)]
A'LENGTH[(N)]	V	array	UI	UI	Number of values in index N of array A
A'ASCENDING[(N)]	V	array	UI	BOOLEAN	True if index N of A is ascending[4]
E'SIMPLE_NAME	V	name		STRING	Simple name of E[4]
E'INSTANCE_NAME	V	name		STRING	Path includes instantiated entities[4]
E'PATH_NAME	V	name		STRING	Path excludes instantiated entities[4]

[1]T=Type, F=Function, V=Value, R=Range.
[2]any=any type or subtype, scalar=scalar type or subtype, discrete=discrete or physical type or subtype, name=entity name=identifier, character literal, or operator symbol.
[3]base(T)=base type of T, T=type of T, UI= universal_integer, T(Result)=type of object described in result column.
[4]Only available in VHDL-93. For 'ASCENDING all enumeration types are ascending.
[5]Or reverse for descending ranges.

TABLE 10.15 Predefined attributes for signals.

Attribute	Kind[1]	Parameter T[2]	Result type[3]	Result/restrictions
S'DELAYED [(T)]	S	TIME	base(S)	S delayed by time T
S'STABLE [(T)]	S	TIME	BOOLEAN	TRUE if no event on S for time T
S'QUIET [(T)]	S	TIME	BOOLEAN	TRUE if S is quiet for time T
S'TRANSACTION	S		BIT	Toggles each cycle if S becomes active
S'EVENT	F		BOOLEAN	TRUE when event occurs on S
S'ACTIVE	F		BOOLEAN	TRUE if S is active
S'LAST_EVENT	F		TIME	Elapsed time since the last event on S
S'LAST_ACTIVE	F		TIME	Elapsed time since S was active
S'LAST_VALUE	F		base(S)	Previous value of S, before last event[4]
S'DRIVING	F		BOOLEAN	TRUE if every element of S is driven[5]
S'DRIVING_VALUE	F		base(S)	Value of the driver for S in the current process[5]

[1] F=function, S=signal.
[2] Time T≥0 ns. The default, if T is not present, is T=0 ns.
[3] base(S)=base type of S.
[4] VHDL-93 returns last value of each signal in array separately as an aggregate, VHDL-87 returns the last value of the composite signal.
[5] VHDL-93 only.

Sequential statements may only appear in processes and subprograms. In the following sections I shall describe each of these different types of sequential statements in turn.

10.10.1 Wait Statement

The **wait statement** is central to VHDL, here are the BNF definitions [VHDL 93LRM8.1]:

```
wait_statement ::= [label:] wait [sensitivity_clause]          [10.24]
  [condition_clause] [timeout_clause] ;
sensitivity_clause ::= on sensitivity_list
sensitivity_list ::= signal_name { , signal_name }
condition_clause ::= until condition
condition ::= boolean_expression
timeout_clause ::= for time_expression
```

A wait statement **suspends** (stops) a process or procedure (you cannot use a wait statement in a function). The wait statement may be made sensitive to events (changes) on **static** signals (the value of the signal must be known at analysis time) that appear in the **sensitivity list** after the keyword on. These signals form the **sensitivity set** of a wait statement. The process will **resume** (restart) when an event occurs on any signal (and only signals) in the sensitivity set.

A wait statement may also contain a condition to be met before the process resumes. If there is no **sensitivity clause** (there is no keyword on) the sensitivity set is made from signals (and only signals) from the **condition clause** that appears after the keyword until (the rules are quite complicated [VHDL 93LRM8.1]).

Finally a wait statement may also contain a **timeout** (following the keyword for) after which the process will resume. Here is the expanded BNF definition, which makes the structure of the wait statement easier to see (but we lose the definitions of the clauses and the sensitivity list):

```
wait_statement ::= [label:] wait
  [on signal_name {, signal_name}]
  [until boolean_expression]
  [for time_expression] ;
```

For example, the statement, **wait on** light, makes you wait until a traffic light changes (any change). The statement, **wait until** light = green, makes you wait (even at a green light) until the traffic signal changes to green. The statement,

```
if light = (red or yellow) then wait until light = green; end if;
```

accurately describes the basic rules at a traffic intersection.

The most common use of the wait statement is to describe synchronous logic, as in the following model of a D flip-flop:

```
entity DFF is port (CLK, D : BIT; Q : out BIT); end;              --1
architecture Behave of DFF is                                     --2
process begin wait until Clk = '1'; Q <= D ; end process;         --3
end;                                                              --4
```

Notice that the statement in line 3 above, **wait until** Clk = '1', is equivalent to **wait on** Clk **until** Clk = '1', and detects a clock edge and not the clock level. Here are some more complex examples of the use of the wait statement:

```
entity Wait_1 is port (Clk, s1, s2 :in BIT); end;
architecture Behave of Wait_1 is
signal x : BIT_VECTOR (0 to 15);
  begin process variable v : BIT; begin
  wait;                          -- Wait forever, stops simulation.
  wait on s1 until s2 = '1'; -- Legal, but s1, s2 are signals so
  -- s1 is in sensitivity list, and s2 is not in the sensitivity set.
  -- Sensitivity set is s1 and process will not resume at event on s2.
  wait on s1, s2;                -- resumes at event on signal s1 or s2.
  wait on s1 for 10 ns;          -- resumes at event on s1 or after 10 ns.
  wait on x;                     -- resumes when any element of array x
                                 -- has an event.
```

```
-- wait on x(1 to v); -- Illegal, nonstatic name, since v is a variable.
end process;
end;

entity Wait_2 is port (Clk, s1, s2:in BIT); end;
architecture Behave of Wait_2 is
  begin process variable v : BIT; begin
  wait on Clk; -- resumes when Clk has an event: rising or falling.
  wait until Clk = '1';          -- resumes on rising edge.
  wait on Clk until Clk = '1';   -- equivalent to the last statement.
  wait on Clk until v = '1';
  -- The above is legal, but v is a variable so
  -- Clk is in sensitivity list, v is not in the sensitivity set.
  -- Sensitivity set is Clk and process will not resume at event on v.
  wait on Clk until s1 = '1';
  -- The above is legal, but s1 is a signal so
  -- Clk is in sensitivity list, s1 is not in the sensitivity set.
  -- Sensitivity set is Clk, process will not resume at event on s1.
  end process;
end;
```

You may only use interface signals that may be read (port modes in, inout, and buffer—see Section 10.7) in the sensitivity list of a wait statement.

10.10.2 Assertion and Report Statements

You can use an **assertion statement** to conditionally issue warnings. The **report statement** (VHDL-93 only) prints an expression and is useful for debugging.

```
assertion_statement ::= [label:] assert                          [10.25]
boolean_expression [report expression] [severity expression] ;

report_statement ::=
[label:] report expression [severity expression] ;
```

Here is an example of an assertion statement:

```
entity Assert_1 is port (I:INTEGER:=0); end;
architecture Behave of Assert_1 is
  begin process begin
  assert (I > 0) report "I is negative or zero"; wait;
  end process;
end;
```

The expression after the keyword report must be of type STRING (the default is "Assertion violation" for the assertion statement), and the expression after the keyword severity must be of type SEVERITY_LEVEL (default ERROR for the assertion statement, and NOTE for the report statement) defined in the STANDARD package. The assertion statement prints if the assertion condition (after the keyword assert) is FALSE. Simulation normally halts for severity of ERROR or FAILURE (you can normally control this threshold in the simulator).

10.10.3 Assignment Statements

There are two sorts of VHDL assignment statements: one for signals and one for variables [VHDL 93LRM8.4–8.5]. The difference is in the timing of the update of the LHS. A **variable assignment statement** is the closest equivalent to the assignment statement in a computer programming language. Variable assignment statements are always sequential statements and the LHS of a variable assignment statement is always updated immediately. Here is the definition and an example:

```
variable_assignment_statement ::=                              [10.26]
    [label:] name|aggregate := expression ;
```

```
entity Var_Assignment is end;
architecture Behave of Var_Assignment is
    signal s1 : INTEGER := 0;
    begin process variable v1,v2 : INTEGER := 0; begin
    assert (v1/=0) report "v1 is 0" severity note ; -- this prints
    v1 := v1 + 1; -- after this statement v1 is 1
    assert (v1=0) report "v1 isn't 0" severity note ; -- this prints
    v2 := v2 + s1; -- signal and variable types must match
    wait;
    end process;
end;
```

This is the output from Cadence Leapfrog for the preceding example:

```
ASSERT/NOTE (time 0 FS) from :$PROCESS_000 (design unit
WORK.VAR_ASSIGNMENT:BEHAVE) v1 is 0
ASSERT/NOTE (time 0 FS) from :$PROCESS_000 (design unit
WORK.VAR_ASSIGNMENT:BEHAVE) v1 isn't 0
```

A **signal assignment statement** schedules a future assignment to a signal:

```
signal_assignment_statement::=                                 [10.27]
    [label:] target <=
    [transport | [ reject time_expression ] inertial] waveform ;
```

The following example shows that, even with no delay, a signal is updated at the end of a simulation cycle after all the other assignments have been scheduled, just before simulation time is advanced:

```
entity Sig_Assignment_1 is end;
architecture Behave of Sig_Assignment_1 is
    signal s1,s2,s3 : INTEGER := 0;
    begin process variable v1 : INTEGER := 1; begin
    assert (s1 /= 0) report "s1 is 0" severity note ; -- this prints.
```

```
  s1 <= s1 + 1; -- after this statement s1 is still 0.
  assert (s1 /= 0) report "s1 still 0" severity note ; -- this prints.
  wait;
  end process;
end;
```

```
ASSERT/NOTE (time 0 FS) from :$PROCESS_000 (design unit
WORK.SIG_ASSIGNMENT_1:BEHAVE) s1 is 0
ASSERT/NOTE (time 0 FS) from :$PROCESS_000 (design unit
WORK.SIG_ASSIGNMENT_1:BEHAVE) s1 still 0
```

Here is an another example to illustrate how time is handled:

```
entity Sig_Assignment_2 is end;
architecture Behave of Sig_Assignment_2 is
  signal s1, s2, s3 : INTEGER := 0;
  begin process variable v1 : INTEGER := 1; begin
  -- s1, s2, s3 are initially 0; now consider the following:
  s1 <= 1 ; -- schedules updates to s1 at end of 0 ns cycle.
  s2 <= s1; -- s2 is 0, not 1.
  wait for 1 ns;
  s3 <= s1; -- now s3 will be 1 at 1 ns.
  wait;
  end process;
end;
```

The Compass simulator produces the following trace file for this example:

```
    Time(fs) + Cycle          s1           s2           s3
------------------------   ----------   ----------   ----------
              0+ 0:           0            0            0
              0+ 1: *         1 *          0            0
...
      1000000+ 1:             1            0 *          1
```

Time is indicated in femtoseconds for each **simulation cycle** plus the number of **delta cycles** (we call this **delta time**, measured in units of **delta**, δ) needed to calculate all transactions on signals. A transaction consists of a new value for a signal (which may be the same as the old value) and the time delay for the value to take effect. An asterisk '*' before a value in the preceding trace indicates that a transaction has occurred and the corresponding signal updated at that time. A transaction that does result in a change in value is an **event**. In the preceding simulation trace for Sig_Assignment_2:Behave

- At 0 ns+ 0δ: all signals are 0.
- At 0 ns+ 1δ: s1 is updated to 1, s2 is updated to 0 (not to 1).
- At 1 ns+ 1δ: s3 is updated to a 1.

The following example shows the behavior of the different **delay models**: **transport** and **inertial** (the default):

```
entity Transport_1 is end;
architecture Behave of Transport_1 is
signal s1, SLOW, FAST, WIRE : BIT := '0';
  begin process begin
  s1 <= '1' after 1 ns, '0' after 2 ns, '1' after 3 ns ;
  -- schedules s1 to be '1' at t+1 ns, '0' at t+2 ns,'1' at t+3 ns
  wait; end process;
-- inertial delay: SLOW rejects pulsewidths less than 5ns:
process (s1) begin SLOW <= s1 after 5 ns ; end process;
-- inertial delay: FAST rejects pulsewidths less than 0.5ns:
process (s1) begin FAST <= s1 after 0.5 ns ; end process;
-- transport delay: WIRE passes all pulsewidths...
process (s1) begin WIRE <= transport s1 after 5 ns ; end process;
end;
```

Here is the trace file from the Compass simulator:

```
        Time(fs) + Cycle    s1 slow fast wire
---------------------       ---- ---- ---- ----
               0+ 0:    '0'  '0'  '0'  '0'
          500000+ 0:    '0'  '0' *'0'  '0'
         1000000+ 0: *'1'  '0'  '0'  '0'
         1500000+ 0:   '1'  '0' *'1'  '0'
         2000000+ 0: *'0'  '0'  '1'  '0'
         2500000+ 0:   '0'  '0' *'0'  '0'
         3000000+ 0: *'1'  '0'  '0'  '0'
         3500000+ 0:   '1'  '0' *'1'  '0'
         5000000+ 0:   '1'  '0'  '1' *'0'
         6000000+ 0:   '1'  '0'  '1' *'1'
         7000000+ 0:   '1'  '0'  '1' *'0'
         8000000+ 0:   '1' *'1'  '1' *'1'
```

Inertial delay mimics the behavior of real logic gates, whereas transport delay more closely models the behavior of wires. In VHDL-93 you can also add a separate **pulse rejection limit** for the inertial delay model as in the following example:

```
process (s1) begin RJCT <= reject 2 ns s1 after 5 ns ; end process;
```

10.10.4 Procedure Call

A **procedure call** in VHDL corresponds to calling a subroutine in a conventional programming language [VHDL LRM8.6]. The parameters in a procedure call statement are the actual procedure parameters (or actuals); the parameters in the procedure definition are the formal procedure parameters (or formals). The two are linked using an association list, which may use either positional or named association (association works just as it does for ports—see Section 10.7.1):

```
procedure_call_statement ::=                              [10.28]
    [label:] procedure_name [(parameter_association_list)];
```

Here is an example:

```
package And_Pkg is
   procedure V_And(a, b : BIT; signal c : out BIT);
   function V_And(a, b : BIT) return BIT;
end;

package body And_Pkg is
   procedure V_And(a, b : BIT; signal c: out BIT) is
      begin c <= a and b; end;
   function V_And(a, b: BIT) return BIT is
      begin return a and b; end;
end And_Pkg;

use work.And_Pkg.all; entity Proc_Call_1 is end;
architecture Behave of Proc_Call_1 is signal A, B, Y: BIT := '0';
   begin process begin V_And (A, B, Y); wait; end process;
end;
```

Table 10.13 on page 416 explains the rules for formal procedure parameters. There is one other way to call procedures, which we shall cover in Section 10.13.3.

10.10.5 If Statement

An **if statement** evaluates one or more Boolean expressions and conditionally executes a corresponding sequence of statements [VHDL LRM8.7].

```
if_statement ::=                                          [10.29]
    [if_label:] if boolean_expression then {sequential_statement}
    {elsif boolean_expression then {sequential_statement}}
    [else {sequential_statement}]
  end if [if_label];
```

The simplest form of an `if` statement is thus:

```
if boolean_expression then {sequential_statement} end if;
```

Here are some examples of the `if` statement:

```
entity If_Then_Else_1 is end;
architecture Behave of If_Then_Else_1 is signal a, b, c: BIT :='1';
   begin process begin
      if c = '1' then c <= a ; else c <= b; end if; wait;
   end process;
end;
```

```
entity If_Then_1 is end;
architecture Behave of If_Then_1 is signal A, B, Y : BIT :='1';
   begin process begin
      if A = B then Y <= A; end if; wait;
   end process;
end;
```

10.10.6 Case Statement

A **case statement** [VHDL LRM8.8] is a multiway decision statement that selects a sequence of statements by matching an expression with a list of (locally static [VHDL LRM7.4.1]) choices.

```
case_statement ::=                                                    [10.30]
[case_label:] case expression is
      when choice {| choice} => {sequential_statement}
   {when choice {| choice} => {sequential_statement}}
end case [case_label];
```

Case statements are useful to model state machines. Here is an example of a Mealy state machine with an asynchronous reset:

```
library IEEE; use IEEE.STD_LOGIC_1164.all;                            --1
entity sm_mealy is                                                    --2
   port (reset, clock, i1, i2 : STD_LOGIC; o1, o2 : out STD_LOGIC);   --3
end sm_mealy;                                                         --4
architecture Behave of sm_mealy is                                    --5
type STATES is (s0, s1, s2, s3); signal current, new : STATES;        --6
begin                                                                 --7
synchronous : process (clock, reset) begin                           --8
   if To_X01(reset) = '0' then current <= s0;                         --9
   elsif rising_edge(clock) then current <= new; end if;              --10
end process;                                                          --11
combinational : process (current, i1, i2) begin                       --12
case current is                                                       --13
   when s0 =>                                                         --14
      if To_X01(i1) = '1' then o2 <='0'; o1 <='0'; new <= s2;         --15
      else o2 <= '1'; o1 <= '1'; new <= s1; end if;                   --16
   when s1 =>                                                         --17
      if To_X01(i2) = '1' then o2 <='1'; o1 <='0'; new <= s1;         --18
      else o2 <='0'; o1 <='1'; new <= s3; end if;                     --19
   when s2 =>                                                         --20
      if To_X01(i2) = '1' then o2 <='0'; o1 <='1'; new <= s2;         --21
      else o2 <= '1'; o1 <= '0'; new <= s0; end if;                   --22
   when s3 => o2 <= '0'; o1 <= '0'; new <= s0;                        --23
   when others => o2 <= '0'; o1 <= '0'; new <= s0;                    --24
end case;                                                             --25
end process;                                                          --26
end Behave;                                                           --27
```

Each possible value of the case expression must be present once, and once only, in the list of choices (or arms) of the case statement (the list must be **exhaustive**). You can use '|' (that means 'or') or 'to' to denote a range in the expression for choice. You may also use the keyword others as the last, default choice (even if the list is already exhaustive, as in the preceding example).

10.10.7 Other Sequential Control Statements

A **loop statement** repeats execution of a series of sequential statements [VHDL LRM8.9]:

```
loop_statement ::=                                          [10.31]
[loop_label:]
[while boolean_expression|for identifier in discrete_range]
loop
   {sequential_statement}
end loop [loop_label];
```

If the **loop variable** (after the keyword for) is used, it is only visible inside the loop. A while loop evaluates the Boolean expression before each execution of the sequence of statements; if the expression is TRUE, the statements are executed. In a for loop the sequence of statements is executed once for each value of the discrete range.

```
package And_Pkg is function V_And(a, b : BIT) return BIT; end;

package body And_Pkg is function V_And(a, b : BIT) return BIT is
   begin return a and b; end; end And_Pkg;

entity Loop_1 is port (x, y : in BIT := '1'; s : out BIT := '0'); end;
use work.And_Pkg.all;
architecture Behave of Loop_1 is
   begin loop
      s <= V_And(x, y); wait on x, y;
   end loop;
end;
```

The **next statement** [VHDL LRM8.10] forces completion of the current iteration of a loop (the containing loop unless another loop label is specified). Completion is forced if the condition following the keyword then is TRUE (or if there is no condition).

```
next_statement ::=                                          [10.32]
[label:] next [loop_label] [when boolean_expression];
```

An **exit statement** [VHDL LRM8.11] forces an exit from a loop.

```
exit_statement ::=
  [label:] exit [loop_label] [when condition] ;
```
[10.33]

As an example:

```
loop wait on Clk; exit when Clk = '0'; end loop;
-- equivalent to: wait until Clk = '0';
```

The **return statement** [VHDL LRM8.12] completes execution of a procedure or function.

```
return_statement ::= [label:] return [expression];
```
[10.34]

A **null statement** [VHDL LRM8.13] does nothing (but is useful in a case statement where all choices must be covered, but for some of the choices you do not want to do anything).

```
null_statement ::= [label:] null;
```
[10.35]

10.11 Operators

Table 10.16 shows the predefined VHDL **operators**, listed by their (increasing) order of precedence [VHDL 93LRM7.2]. The shift operators and the xnor operator were added in VHDL-93.

TABLE 10.16   VHDL predefined operators (listed by increasing order of precedence). [1]

`logical_operator`[2] `::=`	`and | or | nand | nor | xor | xnor`
`relational_operator ::=`	`= | /= | < | <= | > | >=`
`shift_operator`[2] `::=`	`sll | srl | sla | sra | rol | ror`
`adding_operator ::=`	`+ | − | &`
`sign ::=`	`+ | −`
`multiplying_operator ::=`	`* | / | mod | rem`
`miscellaneous_operator ::=`	`** | abs | not`

[1]The not operator is a logical operator but has the precedence of a miscellaneous operator.
[2]Underline means "new to VHDL-93."

The binary **logical operators** (and, or, nand, nor, xor, xnor) and the unary not logical operator are predefined for types BIT or BOOLEAN and one-dimensional arrays whose element type is BIT or BOOLEAN. The operands must be of the same base type for the binary logical operators and the same length if they are arrays.

Both operands of **relational operators** must be of the same type and the result type is BOOLEAN. The equality operator and inequality operator ('=' and '/=') are defined for all types (other than file types). The remaining relational operators, ordering operators, are predefined for any scalar type, and for any one-dimensional array whose elements are of a discrete type (enumeration or integer type).

The left operand of the **shift operators** (VHDL-93 only) is a one-dimensional array with element type of BIT or BOOLEAN; the right operand must be INTEGER.

The **adding operators** ('+' and '−') are predefined for any numeric type. You cannot use the adding operators on BIT or BIT_VECTOR without overloading. The **concatenation operator** '&' is predefined for any one-dimensional array type. The **signs** ('+' and '−') are defined for any numeric type.

The **multiplying operators** are: '*', '/', mod, and rem. The operators '*' and '/' are predefined for any integer or floating-point type, and the operands and the result are of the same type. The operators mod and rem are predefined for any integer type, and the operands and the result are of the same type. In addition, you can multiply an INTEGER or REAL by any physical type and the result is the physical type. You can also divide a physical type by REAL or INTEGER and the result is the physical type. If you divide a physical type by the same physical type, the result is an INTEGER (actually type UNIVERSAL_INTEGER, which is a predefined anonymous type [VHDL LRM7.5]). Once again—you cannot use the multiplying operators on BIT or BIT_VECTOR types without overloading the operators.

The **exponentiating operator**, '**', is predefined for integer and floating-point types. The right operand, the exponent, is type INTEGER. You can only use a negative exponent with a left operand that is a floating-point type, and the result is the same type as the left operand. The unary operator abs (**absolute value**) is predefined for any numeric type and the result is the same type. The operators abs, '**', and not are grouped as **miscellaneous operators**.

Here are some examples of the use of VHDL operators:

```
entity Operator_1 is end; architecture Behave of Operator_1 is        --1
begin process                                                         --2
variable b : BOOLEAN; variable bt : BIT := '1'; variable i : INTEGER; --3
variable pi : REAL := 3.14; variable epsilon : REAL := 0.01;          --4
variable bv4 : BIT_VECTOR (3 downto 0) := "0001";                     --5
variable bv8 : BIT_VECTOR (0 to 7);                                   --6
begin                                                                 --7

b   := "0000" < bv4;  -- b is TRUE, "0000" treated as BIT_VECTOR.     --8
b   := 'f' > 'g';     -- b is FALSE, 'dictionary' comparison.         --9
bt  := '0' and bt;    -- bt is '0', analyzer knows '0' is BIT.        --10
bv4 := not bv4;       -- bv4 is now "1110".                           --11
i   := 1 + 2;         -- Addition, must be compatible types.          --12
```

```
i   := 2 ** 3;        -- Exponentiation, exponent must be integer. --13
i   := 7/3;           -- Division, L/R rounded towards zero, i=2.  --14
i   := 12 rem 7;      -- Remainder, i=5. In general:               --15
                      -- L rem R = L-((L/R)*R).                     --16
i   := 12 mod 7;      -- modulus, i=5. In general:                 --17
                      -- L mod R = L-(R*N) for an integer N.        --18

-- shift := sll | srl | sla | sra | rol | ror (VHDL-93 only)       --19
bv4 := "1001" srl 2;  -- Shift right logical, now bv4="0100".      --20
-- Logical shift fills with T'LEFT.                                --21
bv4 := "1001" sra 2;  -- Shift right arithmetic, now bv4="0111".   --22
-- Arithmetic shift fills with element at end being vacated.       --23
bv4 := "1001" ror 2;  -- Rotate right, now bv4="0110".             --24
-- Rotate wraps around.                                            --25
-- Integer argument to any shift operator may be negative or zero. --26

if (pi*2.718)/2.718 = 3.14 then wait; end if; -- This is unreliable.--27
if (abs(((pi*2.718)/2.718)-3.14)<epsilon) then wait; end if; -- Better.--28

bv8 := bv8(1 to 7) & bv8(0); -- Concatenation, a left rotation.    --29

wait; end process;                                                --30
end;                                                              --31
```

10.12 Arithmetic

The following example illustrates **type checking** and **type conversion** in VHDL arithmetic operations [VHDL 93LRM7.3.4–7.3.5]:

```
entity Arithmetic_1 is end; architecture Behave of Arithmetic_1 is --1
  begin process
    variable i : INTEGER := 1; variable r : REAL := 3.33;         --2
    variable b : BIT := '1';                                      --3
    variable bv4 : BIT_VECTOR (3 downto 0) := "0001";             --4
    variable bv8 : BIT_VECTOR (7 downto 0) := B"1000_0000";       --5
  begin                                                           --6

--   i := r;            -- you can't assign REAL to INTEGER.      --7
--   bv4 := bv4 + 2;    -- you can't add BIT_VECTOR and INTEGER.  --8
--   bv4 := '1';        -- you can't assign BIT to BIT_VECTOR.    --9
--   bv8 := bv4;        -- an error, the arrays are different sizes.--10

r   := REAL(i);        -- OK, uses a type conversion.             --11
i   := INTEGER(r);     -- OK (0.5 rounds up or down).             --12
bv4 := "001" & '1';    -- OK, you can mix an array and a scalar.  --13
bv8 := "0001" & bv4;   -- OK, if arguments are the correct lengths.--14
wait; end process; end;                                          --15
```

The next example shows arithmetic operations between types and subtypes, and also illustrates **range checking** during analysis and simulation:

```
entity Arithmetic_2 is end; architecture Behave of Arithmetic_2 is    --1
type TC is range 0 to 100;              -- Type INTEGER.               --2
type TF is range 32 to 212;             -- Type INTEGER.               --3
subtype STC is INTEGER range 0 to 100;  -- Subtype of type INTEGER.    --4
subtype STF is INTEGER range 32 to 212; -- Base type is INTEGER.       --5
begin process                                                          --6
variable t1 : TC := 25;    variable t2 : TF := 32;                     --7
variable st1 : STC := 25; variable st2 : STF := 32;                    --8
begin                                                                  --9
--    t1    := t2;         -- Illegal, different types.                --10
--    t1    := st1;        -- Illegal, different types and subtypes.   --11
      st2   := st1;        -- OK to use same base types.               --12
      st2   := st1 + 1;    -- OK to use subtype and base type.         --13
--    st2   := 213;        -- Error, outside range at analysis time.   --14
--    st2   := 212 + 1;    -- Error, outside range at analysis time.   --15
      st1   := st1 + 100;  -- Error, outside range at initialization.  --16
wait; end process; end;
```

The MTI simulator, for example, gives the following informative error message during simulation of the preceding model:

```
# ** Fatal: Value 25 is out of range 32 to 212
#    Time: 0 ns  Iteration: 0  Instance:/
# Stopped at Arithmetic_2.vhd line 12
# Fatal error at Arithmetic_2.vhd line 12
```

The assignment st2 := st1 causes this error (since st1 is initialized to 25).

Operations between array types and subtypes are a little more complicated as the following example illustrates:

```
entity Arithmetic_3 is end; architecture Behave of Arithmetic_3 is    --1
type TYPE_1 is array (INTEGER range 3 downto 0) of BIT;               --2
type TYPE_2 is array (INTEGER range 3 downto 0) of BIT;               --3
subtype SUBTYPE_1 is BIT_VECTOR (3 downto 0);                         --4
subtype SUBTYPE_2 is BIT_VECTOR (3 downto 0);                         --5
begin process                                                         --6
variable bv4 : BIT_VECTOR (3 downto 0) := "0001";                     --7
variable st1 : SUBTYPE_1 := "0001"; variable t1 : TYPE_1 := "0001";   --8
variable st2 : SUBTYPE_2 := "0001"; variable t2 : TYPE_2 := "0001";   --9
begin                                                                 --10
      bv4 := st1;          -- OK, compatible type and subtype.        --11
--    bv4 := t1;           -- Illegal, different types.               --12
      bv4 := BIT_VECTOR(t1);  -- OK, type conversion.                 --13
      st1 := bv4;          -- OK, compatible subtype and base type.   --14
--    st1 := t1;           -- Illegal, different types.               --15
      st1 := SUBTYPE_1(t1);   -- OK, type conversion.                 --16
```

```
--    t1    := st1;            -- Illegal, different types.     --17
--    t1    := bv4;            -- Illegal, different types.     --18
      t1    := TYPE_1(bv4);    -- OK, type conversion.          --19
--    t1    := t2;             -- Illegal, different types.     --20
      t1    := TYPE_1(t2);     -- OK, type conversion.          --21
      st1   := st2;            -- OK, compatible subtypes.      --22
wait; end process; end;                                         --23
```

The preceding example uses BIT and BIT_VECTOR types, but exactly the same considerations apply to STD_LOGIC and STD_LOGIC_VECTOR types or other arrays. Notice the use of **type conversion**, written as type_mark'(expression), to convert between **closely related types**. Two types are closely related if they are abstract numeric types (integer or floating-point) or arrays with the same dimension, each index type is the same (or are themselves closely related), and each element has the same type [VHDL 93LRM7.3.5].

10.12.1 IEEE Synthesis Packages

The IEEE 1076.3 standard synthesis packages allow you to perform arithmetic on arrays of the type BIT and STD_LOGIC.[3] The NUMERIC_BIT package defines all of the operators in Table 10.16 (except for the exponentiating operator '**') for arrays of type BIT. Here is part of the package header, showing the declaration of the two types UNSIGNED and SIGNED, and an example of one of the function declarations that overloads the addition operator '+' for UNSIGNED arguments:

```
package Part_NUMERIC_BIT is
type UNSIGNED is array (NATURAL range <> ) of BIT;
type SIGNED is array (NATURAL range <> ) of BIT;
function "+" (L, R : UNSIGNED) return UNSIGNED;
-- other function definitions that overload +, -, = , >, and so on.
end Part_NUMERIC_BIT;
```

The package bodies included in the 1076.3 standard define the functionality of the packages. Companies may implement the functions in any way they wish—as long as the results are the same as those defined by the standard. Here is an example of the parts of the NUMERIC_BIT package body that overload the addition operator '+' for two arguments of type UNSIGNED (even with my added comments the code is rather dense and terse, but remember this is code that we normally never see or need to understand):

```
package body Part_NUMERIC_BIT is
constant NAU : UNSIGNED(0 downto 1) := (others =>'0'); -- Null array.
```

---

[3]IEEE Std 1076.3-1997 was approved by the IEEE Standards Board on 20 March 1997. The synthesis package code on the following pages is reprinted with permission from IEEE Std 1076.3-1997, Copyright © 1997 IEEE. All rights reserved.

```vhdl
constant NAS : SIGNED(0 downto 1):=(others => '0'); -- Null array.
constant NO_WARNING : BOOLEAN := FALSE; -- Default to emit warnings.

function MAX (LEFT, RIGHT : INTEGER) return INTEGER is
begin -- Internal function used to find longest of two inputs.
if LEFT > RIGHT then return LEFT; else return RIGHT; end if; end MAX;

function ADD_UNSIGNED (L, R : UNSIGNED; C: BIT) return UNSIGNED is
constant L_LEFT : INTEGER := L'LENGTH-1; -- L, R must be same length.
alias XL : UNSIGNED(L_LEFT downto 0) is L; -- Descending alias,
alias XR : UNSIGNED(L_LEFT downto 0) is R; -- aligns left ends.
variable RESULT : UNSIGNED(L_LEFT downto 0); variable CBIT : BIT := C;
begin for I in 0 to L_LEFT loop -- Descending alias allows loop.
RESULT(I) := CBIT xor XL(I) xor XR(I); -- CBIT = carry, initially = C.
CBIT := (CBIT and XL(I)) or (CBIT and XR(I)) or (XL(I) and XR(I));
end loop; return RESULT; end ADD_UNSIGNED;

function RESIZE (ARG : UNSIGNED; NEW_SIZE : NATURAL) return UNSIGNED is
constant ARG_LEFT : INTEGER := ARG'LENGTH-1;
alias XARG : UNSIGNED(ARG_LEFT downto 0) is ARG; -- Descending range.
variable RESULT : UNSIGNED(NEW_SIZE-1 downto 0) := (others => '0');
begin -- resize the input ARG to length NEW_SIZE
   if (NEW_SIZE < 1) then return NAU; end if; -- Return null array.
   if XARG'LENGTH = 0 then return RESULT; end if; -- Null to empty.
   if (RESULT'LENGTH < ARG'LENGTH) then -- Check lengths.
      RESULT(RESULT'LEFT downto 0) := XARG(RESULT'LEFT downto 0);
   else -- Need to pad the result with some '0's.
      RESULT(RESULT'LEFT downto XARG'LEFT + 1) := (others => '0');
      RESULT(XARG'LEFT downto 0) := XARG;
   end if; return RESULT;
end RESIZE;

function "+" (L, R : UNSIGNED) return UNSIGNED is -- Overloaded '+'.
constant SIZE : NATURAL := MAX(L'LENGTH, R'LENGTH);
begin -- If length of L or R < 1 return a null array.
if ((L'LENGTH < 1) or (R'LENGTH < 1)) then return NAU; end if;
return ADD_UNSIGNED(RESIZE(L, SIZE), RESIZE(R, SIZE), '0'); end "+";

end Part_NUMERIC_BIT;
```

The following conversion functions are also part of the NUMERIC_BIT package:

```vhdl
function TO_INTEGER (ARG : UNSIGNED) return NATURAL;
function TO_INTEGER (ARG : SIGNED) return INTEGER;
function TO_UNSIGNED (ARG, SIZE : NATURAL) return UNSIGNED;
function TO_SIGNED (ARG : INTEGER; SIZE : NATURAL) return SIGNED;
function RESIZE (ARG : SIGNED; NEW_SIZE : NATURAL) return SIGNED;
function RESIZE (ARG : UNSIGNED; NEW_SIZE : NATURAL) return UNSIGNED;
-- set XMAP to convert unknown values, default is 'X'->'0'
function TO_01(S : UNSIGNED; XMAP : STD_LOGIC := '0') return UNSIGNED;
function TO_01(S : SIGNED; XMAP : STD_LOGIC := '0') return SIGNED;
```

The NUMERIC_STD package is almost identical to the NUMERIC_BIT package except that the UNSIGNED and SIGNED types are declared in terms of the STD_LOGIC type from the Std_Logic_1164 package as follows:

```
library IEEE; use IEEE.STD_LOGIC_1164.all;
package Part_NUMERIC_STD is
type UNSIGNED is array (NATURAL range <>) of STD_LOGIC;
type SIGNED is array (NATURAL range <>) of STD_LOGIC;
end Part_NUMERIC_STD;
```

The NUMERIC_STD package body is similar to NUMERIC_BIT with the addition of a comparison function called STD_MATCH, illustrated by the following:

```
-- function STD_MATCH (L, R: T) return BOOLEAN;
-- T = STD_ULOGIC UNSIGNED SIGNED STD_LOGIC_VECTOR STD_ULOGIC_VECTOR
```

The STD_MATCH function uses the following table to compare logic values:

```
type BOOLEAN_TABLE is array(STD_ULOGIC, STD_ULOGIC) of BOOLEAN;
constant MATCH_TABLE : BOOLEAN_TABLE := (

-----------------------------------------------------------------
-- U     X     0     1     Z     W     L     H     -
-----------------------------------------------------------------
(FALSE,FALSE,FALSE,FALSE,FALSE,FALSE,FALSE,FALSE, TRUE), -- | U |
(FALSE,FALSE,FALSE,FALSE,FALSE,FALSE,FALSE,FALSE, TRUE), -- | X |
(FALSE,FALSE, TRUE,FALSE,FALSE,FALSE, TRUE,FALSE, TRUE), -- | 0 |
(FALSE,FALSE,FALSE, TRUE,FALSE,FALSE,FALSE, TRUE, TRUE), -- | 1 |
(FALSE,FALSE,FALSE,FALSE,FALSE,FALSE,FALSE,FALSE, TRUE), -- | Z |
(FALSE,FALSE,FALSE,FALSE,FALSE,FALSE,FALSE,FALSE, TRUE), -- | W |
(FALSE,FALSE, TRUE,FALSE,FALSE,FALSE, TRUE,FALSE, TRUE), -- | L |
(FALSE,FALSE,FALSE, TRUE,FALSE,FALSE,FALSE, TRUE, TRUE), -- | H |
( TRUE, TRUE, TRUE, TRUE, TRUE, TRUE, TRUE, TRUE, TRUE));-- | - |
```

Thus, for example (notice we need type conversions):

```
IM_TRUE = STD_MATCH(STD_LOGIC_VECTOR    ("10HLXWZ-"),
                    STD_LOGIC_VECTOR    ("HL10----"))    -- is TRUE
```

The following code is similar to the first simple example of Section 10.1, but illustrates the use of the Std_Logic_1164 and NUMERIC_STD packages:

```
entity Counter_1 is end;                              --1
   library STD; use STD.TEXTIO.all;                   --2
   library IEEE; use IEEE.STD_LOGIC_1164.all;         --3
use work.NUMERIC_STD.all;                             --4
architecture Behave_2 of Counter_1 is                 --5
   signal Clock : STD_LOGIC := '0';                   --6
   signal Count : UNSIGNED (2 downto 0) := "000";     --7
   begin                                              --8
   process begin                                      --9
      wait for 10 ns; Clock <= not Clock;             --10
```

```
      if (now > 340 ns) then wait;                      --11
      end if;                                            --12
  end process;                                           --13
  process begin                                          --14
     wait until (Clock = '0');                           --15
     if (Count = 7)                                      --16
        then Count <= "000";                             --17
        else Count <= Count + 1;                         --18
     end if;                                             --19
  end process;                                           --20
  process (Count) variable L: LINE; begin write(L, now); --21
  write(L, STRING'(" Count=")); write(L, TO_INTEGER(Count)); --22
  writeline(output, L);                                  --23
  end process;                                           --24
end;                                                     --25
```

The preceding code looks similar to the code in Section 10.1 (and the output is identical), but there is more going on here:

- Line 3 is a library clause and a use clause for the std_logic_1164 package, so you can use the STD_LOGIC type and the NUMERIC_STD package.
- Line 4 is a use clause for NUMERIC_STD package that was previously analyzed into the library work. If the package is instead analyzed into the library IEEE, you would use the name IEEE.NUMERIC_STD.all here. The NUMERIC_STD package allows you to use the type UNSIGNED.
- Line 6 declares Clock to be type STD_LOGIC and initializes it to '0', instead of the default initial value STD_LOGIC'LEFT (which is 'U').
- Line 7 declares Count to be a 3-bit array of type UNSIGNED from NUMERIC_STD and initializes it using a bit-string literal.
- Line 10 uses the overloaded 'not' operator from std_logic_1164.
- Line 15 uses the overloaded '=' operator from std_logic_1164.
- Line 16 uses the overloaded '=' operator from NUMERIC_STD.
- Line 17 requires a bit-string literal, you cannot use Count <= 0 here.
- Line 18 uses the overloaded '+' operator from NUMERIC_STD.
- Line 22 converts Count, type UNSIGNED, to type INTEGER.

10.13 Concurrent Statements

A **concurrent statement** [VHDL LRM9] is one of the following statements:

```
concurrent_statement ::=                                 [10.36]
   block_statement
 | process_statement
```

```
| [ label : ] [ postponed ] procedure_call ;
| [ label : ] [ postponed ] assertion ;
| [ label : ] [ postponed ] conditional_signal_assignment
| [ label : ] [ postponed ] selected_signal_assignment
| component_instantiation_statement
| generate_statement
```

(The presence of the semicolons ';' in some lines and absence in others in the preceding is correct.) The following sections describe each of these statements in turn.

10.13.1   Block Statement

A **block statement** has the following format [VHDL LRM9.1]:

```
block_statement ::=                                            [10.37]
   block_label: block [(guard_expression)] [is]
      [generic (generic_interface_list);
      [generic map (generic_association_list);]]
      [port (port_interface_list);
      [port map (port_association_list);]]
         {block_declarative_item}
         begin
         {concurrent_statement}
   end block [block_label] ;
```

Blocks may have their own ports and generics and may be used to split an architecture into several hierarchical parts (blocks can also be nested). As a very general rule, for the same reason that it is better to split a computer program into separate small modules, it is usually better to split a large architecture into smaller separate entity–architecture pairs rather than several nested blocks.

A block does have a unique feature: It is possible to specify a **guard expression** for a block. This creates a special signal, GUARD, that you can use within the block to control execution [VHDL LRM9.5]. It also allows you to model three-state buses by declaring **guarded signals** (signal kinds **register** and **bus**).

When you make an assignment statement to a signal, you define a **driver** for that signal. If you make assignments to guarded signals in a block, the driver for that signal is turned off, or **disconnected**, when the GUARD signal is FALSE. The use of guarded signals and guarded blocks can become quite complicated, and not all synthesis tools support these VHDL features.

The following example shows two drivers, A and B, on a three-state bus TSTATE, enabled by signals OEA and OEB. The drivers are enabled by declaring a guard expression after the block declaration and using the keyword guarded in the assignment statements. A **disconnect** statement [VHDL LRM5.3] models the driver delay from driving the bus to the high-impedance state (time to "float").

```vhdl
library ieee; use ieee.std_logic_1164.all;
entity bus_drivers is end;

architecture Structure_1 of bus_drivers is
signal TSTATE: STD_LOGIC bus; signal A, B, OEA, OEB : STD_LOGIC:= '0';
begin
process begin OEA <= '1' after 100 ns, '0' after 200 ns;
OEB <= '1' after 300 ns; wait; end process;
B1 : block (OEA = '1')
disconnect all : STD_LOGIC after 5 ns; -- Only needed for float time.
begin TSTATE <= guarded not A after 3 ns; end block;
B2 : block (OEB = '1')
disconnect all : STD_LOGIC after 5 ns; -- Float time = 5 ns.
begin TSTATE <= guarded not B after 3 ns; end block;
end;
```

		1	2	3	4	5	6	7
Time(fs) + Cycle		tstate	a	b	oea	oeb	b1.GUARD	b2.GUARD
0+ 0:		'U'	'0'	'0'	'0'	'0'	FALSE	FALSE
0+ 1:	*	'Z'	'0'	'0'	'0'	'0'	FALSE	FALSE
100000000+ 0:		'Z'	'0'	'0'	*'1'	'0' *	TRUE	FALSE
103000000+ 0:	*	'1'	'0'	'0'	'1'	'0'	TRUE	FALSE
200000000+ 0:		'1'	'0'	'0'	*'0'	'0' *	FALSE	FALSE
200000000+ 1:	*	'Z'	'0'	'0'	'0'	'0'	FALSE	FALSE
300000000+ 0:		'Z'	'0'	'0'	'0'	*'1'	FALSE *	TRUE
303000000+ 0:	*	'1'	'0'	'0'	'0'	'1'	FALSE	TRUE

Notice the creation of implicit guard signals b1.GUARD and b2.GUARD for each guarded block. There is another, equivalent, method that uses the high-impedance value explicitly as in the following example:

```vhdl
architecture Structure_2 of bus_drivers is
signal TSTATE : STD_LOGIC; signal A, B, OEA, OEB : STD_LOGIC := '0';
begin
process begin
OEA <= '1' after 100 ns, '0' after 200 ns; OEB <= '1' after 300 ns;
wait; end process;
process(OEA, OEB, A, B) begin
  if    (OEA = '1') then TSTATE <= not A after 3 ns;
  elsif (OEB = '1') then TSTATE <= not B after 3 ns;
  else TSTATE <= 'Z' after 5 ns;
```

```
    end if;
end process;
end;
```

This last method is more widely used than the first, and what is more important, more widely accepted by synthesis tools. Most synthesis tools are capable of recognizing the value 'z' on the RHS of an assignment statement as a cue to synthesize a three-state driver. It is up to you to make sure that multiple drivers are never enabled simultaneously to cause contention.

10.13.2 Process Statement

A **process statement** has the following format [VHDL LRM9.2]:

```
process_statement ::=                                    [10.38]
[process_label:]
[postponed] process [(signal_name {, signal_name})]
[is] {subprogram_declaration      | subprogram_body
      | type_declaration          | subtype_declaration
      | constant_declaration      | variable_declaration
      | file_declaration          | alias_declaration
      | attribute_declaration     | attribute_specification
      | use_clause
      | group_declaration         | group_template_declaration}
begin
    {sequential_statement}
end [postponed] process [process_label];
```

The following process models a 2:1 MUX (combinational logic):

```
entity Mux_1 is port (i0, i1, sel : in BIT := '0'; y : out BIT); end;
architecture Behave of Mux_1 is
    begin process (i0, i1, sel) begin -- i0, i1, sel = sensitivity set
    case sel is when '0' => y <= i0; when '1' => y <= i1; end case;
end process; end;
```

This process executes whenever an event occurs on any of the signals in the process **sensitivity set** (i0, i1, sel). The execution of a process occurs during a simulation cycle—a delta cycle. Assignment statements to signals may trigger further delta cycles. Time advances when all transactions for the current time step are complete and all signals updated.

The following code models a two-input AND gate (combinational logic):

```
entity And_1 is port (a, b : in BIT := '0'; y : out BIT); end;
architecture Behave of And_1 is
begin process (a, b) begin y <= a and b; end process; end;
```

The next example models a D flip-flop (sequential logic). The process statement is executed whenever there is an event on clk. The if statement updates the output q with the input d on the rising edge of the signal clk. If the if statement condition is false (as it is on the falling edge of clk), then the assignment statement q <= d will not be executed, and q will keep its previous value. The process thus requires the value of q to be stored between successive process executions, and this implies sequential logic.

```
entity FF_1 is port (clk, d: in BIT := '0'; q : out BIT); end;
architecture Behave of FF_1 is
begin process (clk) begin
   if clk'EVENT and clk = '1' then q <= d; end if;
end process; end;
```

The behavior of the next example is identical to the previous model. Notice that the wait statement is at the end of the equivalent process with the signals in the sensitivity set (in this case just one signal, clk) included in the sensitivity list (that follows the keyword on).

```
entity FF_2 is port (clk, d: in BIT := '0'; q : out BIT); end;
architecture Behave of FF_2 is
begin process begin -- The equivalent process has a wait at the end:
    if clk'event and clk = '1' then q <= d; end if; wait on clk;
end process; end;
```

If we use a wait statement in a process statement, then we may not use a process sensitivity set (the reverse is true: If we do not have a sensitivity set for a process, we must include a wait statement or the process will execute endlessly):

```
entity FF_3 is port (clk, d: in BIT := '0'; q : out BIT); end;
architecture Behave of FF_3 is
begin process begin -- No sensitivity set with a wait statement.
   wait until clk = '1'; q <= d;
end process; end;
```

If you include ports (interface signals) in the sensitivity set of a process statement, they must be ports that can be read (they must be of mode in, inout, or buffer, see Section 10.7).

10.13.3 Concurrent Procedure Call

A **concurrent procedure call** appears outside a process statement [VHDL LRM9.3]. The concurrent procedure call is a shorthand way of writing an equivalent process statement that contains a procedure call (Section 10.10.4):

```
package And_Pkg is procedure V_And(a,b:BIT; signal c:out BIT); end;

package body And_Pkg is procedure V_And(a,b:BIT; signal c:out BIT) is
   begin c <= a and b; end; end And_Pkg;

use work.And_Pkg.all; entity Proc_Call_2 is end;
architecture Behave of Proc_Call_2 is signal A, B, Y : BIT := '0';
   begin V_And (A, B, Y); -- Concurrent procedure call.
```

```
process begin wait; end process; -- Extra process to stop.
end;
```

10.13.4 Concurrent Signal Assignment

There are two forms of **concurrent signal assignment statement**. A **selected signal assignment statement** is equivalent to a case statement inside a process statement [VHDL LRM9.5.2]:

```
selected_signal_assignment ::=                                    [10.39]
  with expression select
    name|aggregate <= [guarded]
      [transport|[reject time expression] inertial]
        waveform when choice {| choice}
          {, waveform when choice {| choice} } ;
```

The following design unit, Selected_1, uses a selected signal assignment. The equivalent unit, Selected_2, uses a case statement inside a process statement.

```
entity Selected_1 is end; architecture Behave of Selected_1 is
signal y,i1,i2 : INTEGER; signal sel : INTEGER range 0 to 1;
begin with sel select y <= i1 when 0, i2 when 1; end;
```

```
entity Selected_2 is end; architecture Behave of Selected_2 is
signal i1,i2,y : INTEGER; signal sel : INTEGER range 0 to 1;
begin process begin
  case sel is when 0 => y <= i1; when 1 => y <= i2; end case;
  wait on i1, i2;
end process; end;
```

The other form of concurrent signal assignment is a **conditional signal assignment statement** that, in its most general form, is equivalent to an if statement inside a process statement [VHDL LRM9.5.1]:

```
conditional_signal_assignment ::=                                 [10.40]
    name|aggregate <= [guarded]
  [transport|[reject time expression] inertial]
      {waveform when boolean_expression else}
        waveform [when boolean_expression];
```

Notice that in VHDL-93 the else clause is optional. Here is an example of a conditional signal assignment, followed by a model using the equivalent process with an if statement:

```
entity Conditional_1 is end; architecture Behave of Conditional_1 is
signal y,i,j : INTEGER; signal clk : BIT;
begin y <= i when clk = '1' else j; -- conditional signal assignment
end;
```

```
entity Conditional_2 is end; architecture Behave of Conditional_2 is
signal y,i : INTEGER; signal clk : BIT;
begin process begin
```

```
     if clk = '1' then y <= i; else y <= y ; end if; wait on clk;
end process; end;
```

A concurrent signal assignment statement can look just like a sequential signal assignment statement, as in the following example:

```
entity Assign_1 is end; architecture Behave of Assign_1 is
signal Target, Source : INTEGER;
   begin Target <= Source after 1 ns; -- looks like signal assignment
end;
```

However, outside a `process` statement, this statement is a concurrent signal assignment and has its own equivalent `process` statement. Here is the equivalent process for the example:

```
entity Assign_2 is end; architecture Behave of Assign_2 is
signal Target, Source : INTEGER;
begin process begin
   Target <= Source after 1 ns; wait on Source;
end process; end;
```

Every process is executed once during initialization. In the previous example, an initial value will be scheduled to be assigned to `Target` even though there is no event on `Source`. If, for some reason, you do not want this to happen, you need to rewrite the concurrent assignment statement as a `process` statement with a `wait` statement before the assignment statement:

```
entity Assign_3 is end; architecture Behave of Assign_3 is
signal Target, Source : INTEGER; begin process begin
   wait on Source; Target <= Source after 1 ns;
end process; end;
```

10.13.5 Concurrent Assertion Statement

A **concurrent assertion statement** is equivalent to a passive `process` statement (without a sensitivity list) that contains an `assertion` statement followed by a `wait` statement [VHDL LRM9.4].

```
concurrent_assertion_statement                                    [10.41]
::= [ label : ] [ postponed ] assertion ;
```

If the assertion condition contains a signal, then the equivalent `process` statement will include a final `wait` statement with a sensitivity clause. A concurrent assertion statement with a condition that is static expression is equivalent to a `process` statement that ends in a `wait` statement that has no sensitivity clause. The equivalent process will execute once, at the beginning of simulation, and then wait indefinitely.

10.13.6 Component Instantiation

A **component instantiation statement** in VHDL is similar to placement of a component in a schematic—an instantiated component is somewhere between a copy of the component and a reference to the component. Here is the definition [VHDL LRM9.6]:

```
component_instantiation_statement ::=                          [10.42]
instantiation_label:
 [component] component_name
|entity entity_name [(architecture_identifier)]
|configuration configuration_name
  [generic map (generic_association_list)]
  [port map (port_association_list)] ;
```

We examined component instantiation using a *component*_name in Section 10.5. If we instantiate a component in this way we must declare the component (see BNF [10.9]). To bind a component to an entity–architecture pair we can use a configuration, as illustrated in Figure 10.1, or we can use the default binding as described in Section 10.7. In VHDL-93 we have another alternative—we can directly instantiate an entity or configuration. For example:

```
entity And_2 is port (i1, i2 : in BIT; y : out BIT); end;
architecture Behave of And_2 is begin y <= i1 and i2; end;
entity Xor_2 is port (i1, i2 : in BIT; y : out BIT); end;
architecture Behave of Xor_2 is begin y <= i1 xor i2; end;

entity Half_Adder_2 is port (a,b : BIT := '0'; sum, cry : out BIT); end;
architecture Netlist_2 of Half_Adder_2 is
use work.all; -- need this to see the entities Xor_2 and And_2
begin
   X1 : entity Xor_2(Behave) port map (a, b, sum); -- VHDL-93 only
   A1 : entity And_2(Behave) port map (a, b, cry); -- VHDL-93 only
end;
```

10.13.7 Generate Statement

A **generate statement** [VHDL LRM9.7] simplifies repetitive code:

```
generate_statement ::=                                         [10.43]
generate_label:   for generate_parameter_specification
                 |if boolean_expression
generate [{block_declarative_item} begin]
  {concurrent_statement}
end generate [generate_label] ;
```

Here is an example (notice the labels are required):

```
entity Full_Adder is port (X, Y, Cin : BIT; Cout, Sum: out BIT); end;
architecture Behave of Full_Adder is begin Sum <= X xor Y xor Cin;
Cout <= (X and Y) or (X and Cin) or (Y and Cin); end;
```

```
entity Adder_1 is
  port (A, B : in BIT_VECTOR (7 downto 0) := (others => '0');
  Cin : in BIT := '0'; Sum : out BIT_VECTOR (7 downto 0);
  Cout : out BIT);
end;

architecture Structure of Adder_1 is use work.all;

component Full_Adder port (X, Y, Cin: BIT; Cout, Sum: out BIT);
end component;
signal C : BIT_VECTOR(7 downto 0);
begin AllBits : for i in 7 downto 0 generate
  LowBit : if i = 0 generate
    FA : Full_Adder port map (A(0), B(0), Cin, C(0), Sum(0));
  end generate;
  OtherBits : if i /= 0 generate
    FA : Full_Adder port map (A(i), B(i), C(i-1), C(i), Sum(i));
  end generate;
end generate;
Cout <= C(7);
end;
```

The instance names within a generate loop include the generate parameter.
For example for i=6, FA'INSTANCE_NAME is

```
:adder_1(structure):allbits(6):otherbits:fa:
```

10.14 Execution

Two successive statements may execute in either a concurrent or sequential fashion
depending on where the statements appear.

```
statement_1; statement_2;
```

In **sequential execution,** statement_1 in this sequence is always evaluated before
statement 2. In **concurrent execution,** statement_1 and statement_2 are
evaluated at the same time (as far as we are concerned—obviously on most comput-
ers exactly parallel execution is not possible). Concurrent execution is the most
important difference between VHDL and a computer programming language. Suppose we have two signal assignment statements inside a process statement. In this
case statement_1 and statement_2 are sequential assignment statements:

```
entity Sequential_1 is end; architecture Behave of Sequential_1 is
signal s1, s2 : INTEGER := 0;
begin
  process begin
    s1 <= 1;              -- sequential signal assignment 1
    s2 <= s1 + 1;         -- sequential signal assignment 2
    wait on s1, s2 ;
```

```
    end process;
end;
```

```
    Time(fs) + Cycle              s1           s2
----------------------    ------------ ------------
              0+ 0:              0            0
              0+ 1: *            1 *          1
              0+ 2: *            1 *          2
              0+ 3: *            1 *          2
```

If the two statements are outside a process statement they are concurrent assignment statements, as in the following example:

```
entity Concurrent_1 is end; architecture Behave of Concurrent_1 is
signal s1, s2 : INTEGER := 0; begin
    L1 : s1 <= 1;        -- concurrent signal assignment 1
    L2 : s2 <= s1 + 1;   -- concurrent signal assignment 2
end;
```

```
    Time(fs) + Cycle              s1           s2
----------------------    ------------ ------------
              0+ 0:              0            0
              0+ 1: *            1 *          1
              0+ 2:              1 *          2
```

The two concurrent signal assignment statements in the previous example are equivalent to the two processes, labeled as P1 and P2, in the following model.

```
entity Concurrent_2 is end; architecture Behave of Concurrent_2 is
signal s1, s2 : INTEGER := 0; begin
    P1 : process begin s1 <= 1;        wait on s2 ; end process;
    P2 : process begin s2 <= s1 + 1; wait on s1 ; end process;
end;
```

```
    Time(fs) + Cycle              s1           s2
----------------------    ------------ ------------
              0+ 0:              0            0
              0+ 1: *            1 *          1
              0+ 2: *            1 *          2
              0+ 3: *            1            2
```

Notice that the results are the same (though the trace files are slightly different) for the architectures Sequential_1, Concurrent_1, and Concurrent_2. Updates to signals occur at the end of the simulation cycle, so the values used will always be the old values. So far things seem fairly simple: We have sequential execution or concurrent execution. However, variables are updated immediately, so the variable values that are used are always the new values. The examples in Table 10.17 illustrate this very important difference.

The various concurrent and sequential statements in VHDL are summarized in Table 10.18.

TABLE 10.17   Variables and signals in VHDL.

Variables	Signals
| ```
entity Execute_1 is end;
architecture Behave of Execute_1 is
begin
  process
  variable v1 : INTEGER := 1;
  variable v2 : INTEGER := 2;
  begin
    v1 := v2; -- before: v1 = 1, v2 = 2
    v2 := v1; -- after:  v1 = 2, v2 = 2
    wait;
  end process;
end;
``` | ```
entity Execute_2 is end;
architecture Behave of Execute_2 is
signal s1 : INTEGER := 1;
signal s2 : INTEGER := 2;
begin
  process
  begin
    s1 <= s2; -- before: s1 = 1, s2 = 2
    s2 <= s1; -- after:  s1 = 2, s2 = 1
    wait;
  end process;
end;
``` |

TABLE 10.18   Concurrent and sequential statements in VHDL.

Concurrent [VHDL LRM9]	Sequential [VHDL LRM8]	
block	wait	case
process	assertion	loop
concurrent_procedure_call	signal_assignment	next
concurrent_assertion	variable_assignment	exit
concurrent_signal_assignmen	procedure_call	return
t	if	null
component_instantiation		
generate		

10.15 Configurations and Specifications

The difference between, the interaction, and the use of component/configuration declarations and specifications is probably the most confusing aspect of VHDL. Fortunately this aspect of VHDL is not normally important for ASIC design. The syntax of component/configuration declarations and specifications is shown in Table 10.19.

- A *configuration declaration* defines a configuration—it is a library unit and is one of the basic units of VHDL code.

- A *block configuration* defines the configuration of a block statement or a design entity. A block configuration appears inside a configuration declaration, a component configuration, or nested in another block configuration.

TABLE 10.19 VHDL binding.

configuration declaration[1] [VHDL LRM1.3]	**configuration** identifier **of** *entity*_name **is** {use_clause|attribute_specification|group_declaration} block_configuration **end** [**configuration**] [*configuration*_identifier];
block configuration [VHDL LRM1.3.1]	**for** *architecture*_name |*block_statement*_label |*generate_statement*_label [(index_specification)] {**use** selected_name {, selected_name};} {block_configuration|component_configuration} **end for** ;
configuration specification[1] [VHDL LRM5.2]	**for** *instantiation*_label{,*instantiation*_label}:*component*_name |**others**:*component*_name |**all**:*component*_name [**use** **entity** *entity*_name [(*architecture*_identifier)] |**configuration** *configuration*_name |**open**] [**generic map** (*generic*_association_list)] [**port map** (*port*_association_list)];
component declaration[1] [VHDL LRM4.5]	**component** identifier [**is**] [**generic** (*local_generic*_interface_list);] [**port** (*local_port*_interface_list);] **end component** [*component*_identifier];
component configuration[1] [VHDL LRM1.3.2]	**for** *instantiation*_label {, *instantiation*_label}:*component*_name |**others**:*component*_name |**all**:*component*_name [[**use** **entity** *entity*_name [(*architecture*_identifier)] |**configuration** *configuration*_name |**open**] [**generic map** (*generic*_association_list)] [**port map** (*port*_association_list)];] [block_configuration] **end for**;

[1]Underline means "new to VHDL-93".

- A *configuration specification* may appear in the declarative region of a generate statement, block statement, or architecture body.
- A *component declaration* may appear in the declarative region of a generate statement, block statement, architecture body, or package.
- A *component configuration* defines the configuration of a component and appears in a block configuration.

Table 10.20 shows a simple example (identical in structure to the example of Section 10.5) that illustrates the use of each of the preceding constructs.

TABLE 10.20   VHDL binding examples.

	```entity AD2 is port (A1, A2: in BIT; Y: out BIT); end;```
	```architecture B of AD2 is begin Y <= A1 and A2; end;```
	```entity XR2 is port (X1, X2: in BIT; Y: out BIT); end;```
	```architecture B of XR2 is begin Y <= X1 xor X2; end;```

component
declaration
 configuration
 specification
	```entity Half_Adder is port (X, Y: BIT; Sum, Cout: out BIT); end;```
	```architecture Netlist of Half_Adder is use work.all;```
	```component MX port (A, B: BIT; Z :out BIT);end component;```
	```component MA port (A, B: BIT; Z :out BIT);end component;```
	```for G1:MX use entity XR2(B) port map(X1 => A,X2 => B,Y => Z);```
	```begin```
	   ```G1:MX port map(X, Y, Sum); G2:MA port map(X, Y, Cout);```
	```end;```

configuration
declaration
 block
 configuration
  component
  configuration
	```configuration C1 of Half_Adder is```
	```use work.all;```
	  ```for Netlist```
	    ```for G2:MA```
	      ```use entity AD2(B) port map(A1 => A,A2 => B,Y => Z);```
	    ```end for;```
	  ```end for;```
	```end;```

10.16 An Engine Controller

This section describes part of a controller for an automobile engine. Table 10.21 shows a temperature converter that converts digitized temperature readings from a sensor from degrees Centigrade to degrees Fahrenheit.

To save area the temperature conversion is approximate. Instead of multiplying by 9/5 and adding 32 (so 0°C becomes 32°F and 100°C becomes 212°F) we multiply by 1.75 and add 32 (so 100°C becomes 207°F). Since $1.75 = 1 + 0.5 + 0.25$, we can multiply by 1.75 using shifts (for divide by 2, and divide by 4) together with a very simple constant addition (since 32=`"100000"`). Using shift to multiply and divide by powers of 2 is free in hardware (we just change connections to a bus). For

TABLE 10.21 A temperature converter.

```
library IEEE;
use IEEE.STD_LOGIC_1164.all; -- type STD_LOGIC, rising_edge
use IEEE.NUMERIC_STD.all ; -- type UNSIGNED, "+", "/"
entity tconv is generic TPD : TIME:= 1 ns;
   port (T : in UNSIGNED(11 downto 0);
   T_out : out UNSIGNED(11 downto 0));
end;
architecture rtl of tconv is
constant T2  : UNSIGNED(1 downto 0) := "10" ;
constant T4  : UNSIGNED(2 downto 0) := "100" ;
constant T32 : UNSIGNED(5 downto 0) := "100000" ;
begin
   process(T) begin
      T_out <= T + T/T2 + T/T4 + T32 after TPD;
   end process;
end rtl;
```

T = temperature in °C

T_out = temperature in °F

The conversion formula from Centigrade to Fahrenheit is:
$T(°F) = (9/5) \times T(°C) + 32$

This converter uses the approximation:
$9/5 \approx 1.75 = 1 + 0.5 + 0.25$

large temperatures the error approaches 0.05/1.8 or approximately 3 percent. We play these kinds of tricks often in hardware computation. Notice also that temperatures measured in °C and °F are defined as unsigned integers of the same width. We could have defined these as separate types to take advantage of VHDL's type checking.

Table 10.22 describes a digital filter to compute a "moving average" over four successive samples in time (i(0), i(1), i(2), and i(3)), with i(0) being the first sample).

The filter uses the following formula:

$$T\_out <= (i(0) + i(1) + i(2) + i(3))/T4$$

Division by T4="100" is free in hardware. If instead, we performed the divisions before the additions, this would reduce the number of bits to be added for two of the additions and saves us worrying about overflow. The drawback to this approach is round-off errors. We can use the register shown in Table 10.23 to register the inputs.

Table 10.24 shows a **first-in, first-out** stack (**FIFO**). This allows us to buffer the signals coming from the sensor until the microprocessor has a chance to read them. The depth of the FIFO will depend on the maximum amount of time that can pass without the microcontroller being able to read from the bus. We have to determine this with statistical simulations taking into account other traffic on the bus.

TABLE 10.22   A digital filter.

```
library IEEE;
use IEEE.STD_LOGIC_1164.all; -- STD_LOGIC type, rising_edge
use IEEE.NUMERIC_STD.all; -- UNSIGNED type, "+" and "/"
entity filter is
   generic TPD : TIME := 1 ns;
   port (T_in : in UNSIGNED(11 downto 0);
   rst, clk : in STD_LOGIC;
   T_out: out UNSIGNED(11 downto 0));
end;
architecture rtl of filter is
type arr is array (0 to 3) of UNSIGNED(11 downto 0);
signal i : arr ;
constant T4 : UNSIGNED(2 downto 0) := "100";
begin
   process(rst, clk) begin
      if (rst = '1') then
         for n in 0 to 3 loop i(n) <= (others =>'0') after TPD;
         end loop;
      else
         if(rising_edge(clk)) then
         i(0) <= T_in after TPD;i(1) <= i(0) after TPD;
         i(2) <= i(1) after TPD;i(3) <= i(2) after TPD;
         end if;
      end if;
   end process;
   process(i) begin
      T_out <= ( i(0) + i(1) + i(2) + i(3) )/T4 after TPD;
   end process;
end rtl;
```

The filter computes a moving average over four successive samples in time.

Notice
i(0) i(1) i(2) i(3)
are each 12 bits wide.

Then the sum
i(0) + i(1) + i(2) + i(3)
is 14 bits wide, and the average

(i(0) + i(1) + i(2) + i(3))/T4

is 12 bits wide.

All delays are generic TPD.

The FIFO has flags, empty and full, that signify its state. It uses a function to increment two circular pointers. One pointer keeps track of the address to write to next, the other pointer tracks the address to read from. The FIFO memory may be implemented in a number of ways in hardware. We shall assume for the moment that it will be synthesized as a bank of flip-flops.

Table 10.25 shows a controller for the two FIFOs. The controller handles the reading and writing to the FIFO. The microcontroller attached to the bus signals which of the FIFOs it wishes to read from. The controller then places the appropriate data on the bus. The microcontroller can also ask for the FIFO flags to be placed in the low-order bits of the bus on a read cycle. If none of these actions are requested by the microcontroller, the FIFO controller three-states its output drivers.

TABLE 10.23   The input register.

```
library IEEE;
use IEEE.STD_LOGIC_1164.all; -- type STD_LOGIC, rising_edge
use IEEE.NUMERIC_STD.all ; -- type UNSIGNED
entity register_in is
generic ( TPD : TIME := 1 ns);
port (T_in : in UNSIGNED(11 downto 0);
clk, rst : in STD_LOGIC; T_out : out UNSIGNED(11 downto 0)); end;
architecture rtl of register_in is
begin
  process(clk, rst) begin
    if (rst = '1') then T_out <= (others => '0') after TPD;
    else
    if (rising_edge(clk)) then T_out <= T_in after TPD; end if;
    end if;
  end process;
end rtl ;
```

12-bit-wide register for the temperature input signals.

If the input is asynchronous (from an A/D converter with a separate clock, for example), we would need to worry about metastability.

All delays are generic TPD.

Table 10.25 shows the top level of the controller. To complete our model we shall use a package for the component declarations:

```
package TC_Components is
component register_in generic (TPD : TIME := 1 ns);
port (T_in : in UNSIGNED(11 downto 0);
   clk, rst : in STD_LOGIC; T_out : out UNSIGNED(11 downto 0));
end component;
component tconv generic (TPD : TIME := 1 ns);
port (T : in UNSIGNED (11 downto 0);
   T_out : out UNSIGNED(11 downto 0));
end component;
component filter generic (TPD : TIME := 1 ns);
port (T_in : in UNSIGNED (11 downto 0);
   rst, clk : in STD_LOGIC; T_out : out UNSIGNED(11 downto 0));
end component;
component fifo generic (width:INTEGER := 12; depth : INTEGER := 16);
   port (clk, rst, push, pop  :  STD_LOGIC;
     Di :  UNSIGNED (width-1 downto 0);
     Do : out UNSIGNED (width-1 downto 0);
     empty, full : out STD_LOGIC);
end component;
component fifo_control generic (TPD:TIME := 1 ns);
   port (D_1, D_2 : in UNSIGNED(11 downto 0);
     select : in UNSIGNED(1 downto 0); read, f1, f2, e1, e2 : in STD_LOGIC;
     r1, r2, w12 : out STD_LOGIC; D : out UNSIGNED(11 downto 0)) ;
```

TABLE 10.24   A first-in, first-out stack (FIFO).

```
library IEEE; use IEEE.NUMERIC_STD.all ; -- UNSIGNED type
use ieee.std_logic_1164.all; -- STD_LOGIC type, rising_edge
entity fifo is
    generic (width : INTEGER := 12; depth : INTEGER := 16);
    port (clk, rst, push, pop : STD_LOGIC;
    Di : in UNSIGNED (width-1 downto 0);
    Do : out UNSIGNED (width-1 downto 0);
    empty, full : out STD_LOGIC);
end fifo;
architecture rtl of fifo is
subtype ptype is INTEGER range 0 to (depth-1);
signal diff, Ai, Ao : ptype; signal  f, e : STD_LOGIC;
type a is array (ptype) of UNSIGNED(width-1 downto 0);
signal mem : a ;
function bump(signal ptr : INTEGER range 0 to (depth-1))
return INTEGER is begin
    if (ptr = (depth-1)) then return 0;
    else return (ptr + 1);
    end if;
end;
begin
    process(f,e) begin full <= f ; empty <= e; end process;
    process(diff) begin
    if (diff = depth -1) then f <= '1'; else f <= '0'; end if;
    if (diff = 0) then e <= '1'; else e <= '0'; end if;
    end process;
    process(clk, Ai, Ao, Di, mem, push, pop, e, f) begin
    if(rising_edge(clk)) then
    if(push='0')and(pop='1')and(e='0') then Do <= mem(Ao); end if;
    if(push='1')and(pop='0')and(f='0') then mem(Ai) <= Di; end if;
    end if ;
    end process;
    process(rst, clk)  begin
        if(rst = '1') then Ai <= 0; Ao <= 0; diff <= 0;
        else if(rising_edge(clk)) then
            if (push = '1') and (f = '0') and (pop = '0') then
            Ai <= bump(Ai); diff <= diff + 1;
            elsif (pop = '1') and (e = '0') and (push = '0') then
            Ao <= bump(Ao); diff <= diff - 1;
            end if;
        end if;
        end if;
    end process;
end;
```

FIFO (first-in, first-out) register

Reads (pop = 1) and writes (push = 1) are synchronous to the rising edge of the clock. Read and write should not occur at the same time. The width (number of bits in each word) and depth (number of words) are generics.

External signals:
clk, clock
rst, reset active-high
push, write to FIFO
pop, read from FIFO
Di, data in
Do, data out
empty, FIFO flag
full, FIFO flag

Internal signals:
diff, difference pointer
Ai, input address
Ao, output address
f, full flag
e, empty flag

No delays in this model.

TABLE 10.25 A FIFO controller.

```
library IEEE;use IEEE.STD_LOGIC_1164.all;use IEEE.NUMERIC_STD.all;
entity fifo_control is generic TPD : TIME := 1 ns;
   port(D_1, D_2 : in UNSIGNED(11 downto 0);
   sel : in UNSIGNED(1 downto 0) ;
   read , f1, f2, e1, e2 : in STD_LOGIC;
   r1, r2, w12 : out STD_LOGIC; D : out UNSIGNED(11 downto 0)) ;
end;
architecture rtl of fifo_control is
   begin process
   (read, sel, D_1, D_2, f1, f2, e1, e2)
   begin
   r1 <= '0' after TPD; r2 <= '0' after TPD;
   if (read = '1') then
      w12 <= '0' after TPD;
      case sel is
      when "01" =>   D <= D_1 after TPD; r1 <= '1' after TPD;
      when "10" =>   D <= D_2 after TPD; r2 <= '1' after TPD;
      when "00" =>   D(3) <= f1 after TPD; D(2) <= f2 after TPD;
                     D(1) <= e1 after TPD; D(0) <= e2 after TPD;
      when others => D <= "ZZZZZZZZZZZZ" after TPD;
      end case;
   elsif (read = '0') then
      D <= "ZZZZZZZZZZZZ" after TPD; w12 <= '1' after TPD;
   else D <= "ZZZZZZZZZZZZ" after TPD;
   end if;
   end process;
end rtl;
```

This handles the reading and writing to the FIFOs under control of the processor (mpu). The mpu can ask for data from either FIFO or for status flags to be placed on the bus.

Inputs:
D_1
   data in from FIFO1
D_2
   data in from FIFO2
sel
   FIFO select from mpu
read
   FIFO read from mpu
f1,f2,e1,e2
   flags from FIFOs

Outputs:
r1, r2
   read enables for FIFOs
w12
   write enable for FIFOs
D
   data out to mpu bus

```
         end component;
         end;
```

The following testbench completes a set of reads and writes to the FIFOs:

```
library IEEE;
use IEEE.std_logic_1164.all; -- type STD_LOGIC
use IEEE.numeric_std.all; -- type UNSIGNED
entity test_TC is end;
architecture testbench of test_TC is
component T_Control port (T_1, T_2 : in UNSIGNED(11 downto 0);
   clk : in STD_LOGIC; sensor: in UNSIGNED( 1 downto 0) ;
   read : in STD_LOGIC; rst : in STD_LOGIC;
   D : out UNSIGNED(11 downto 0)); end component;
signal T_1, T_2 : UNSIGNED(11 downto 0);
```

TABLE 10.26   Top level of temperature controller.

```
library IEEE; use IEEE.STD_LOGIC_1164.all; use IEEE.NUMERIC_STD.all;
entity T_Control is port (T_in1, T_in2 : in UNSIGNED (11 downto 0);
   sensor: in UNSIGNED(1 downto 0);
   clk, RD, rst : in STD_LOGIC; D : out UNSIGNED(11 downto 0));
end;
architecture structure of T_Control is use work.TC_Components.all;
signal F, E : UNSIGNED (2 downto 1);
signal T_out1, T_out2, R_out1, R_out2, F1, F2, FIFO1, FIFO2 : UNSIGNED(11 downto 0);
signal RD1, RD2, WR: STD_LOGIC ;
begin
RG1 : register_in generic map (1ns) port map (T_in1, clk, rst, R_out1);
RG2 : register_in generic map (1ns) port map (T_in2, clk, rst, R_out2);
TC1 : tconv generic map (1ns) port map (R_out1, T_out1);
TC2 : tconv generic map (1ns) port map (R_out2, T_out2);
TF1 : filter generic map (1ns) port map (T_out1, rst, clk, F1);
TF2 : filter generic map (1ns) port map (T_out2, rst, clk, F2);
FI1 : fifo generic map (12,16) port map (clk, rst, WR, RD1, F1, FIFO1, E(1), F(1));
FI2 : fifo generic map (12,16) port map (clk, rst, WR, RD2, F2, FIFO2, E(2), F(2));
FC1 : fifo_control port map
(FIFO1, FIFO2, sensor, RD, F(1), F(2), E(1), E(2), RD1, RD2, WR, D);
end structure;
```

```
signal clk, read, rst : STD_LOGIC;
signal sensor : UNSIGNED(1 downto 0);
signal D : UNSIGNED(11 downto 0);
begin TT1 : T_Control port map (T_1, T_2, clk, sensor, read, rst, D);
process begin
rst <= '0'; clk <= '0';
wait for 5 ns; rst <= '1'; wait for 5 ns; rst <= '0';
T_1 <= "000000000011"; T_2 <= "000000000111"; read <= '0';
   for i in 0 to 15 loop -- fill the FIFOs
   clk <= '0'; wait for 5ns; clk <= '1'; wait for 5 ns;
   end loop;
   assert (false) report "FIFOs full" severity NOTE;
   clk <= '0'; wait for 5ns; clk <= '1'; wait for 5 ns;
read <= '1'; sensor <= "01";
   for i in 0 to 15 loop -- empty the FIFOs
   clk <= '0'; wait for 5ns; clk <= '1'; wait for 5 ns;
   end loop;
   assert (false) report "FIFOs empty" severity NOTE;
   clk <= '0'; wait for 5ns; clk <= '1'; wait;
end process;
end;
```

10.17 Summary

Table 10.27 shows the essential elements of the VHDL language. Table 10.28 shows the most important BNF definitions and their locations in this chapter. The key points covered in this chapter are as follows:

- The use of an `entity` and an `architecture`
- The use of a `configuration` to bind entities and their architectures
- The compile, elaboration, initialization, and simulation steps
- Types, subtypes, and their use in expressions
- The logic systems based on `BIT` and `Std_Logic_1164` types
- The use of the IEEE synthesis packages for `BIT` arithmetic
- Ports and port modes
- Initial values and the difference between simulation and hardware
- The difference between a `signal` and a `variable`
- The different assignment statements and the timing of updates
- The `process` and `wait` statements

VHDL is a "wordy" language. The examples in this chapter are complete rather than code fragments. To write VHDL "nicely," with indentation and nesting of constructs, requires a large amount of space. Some of the VHDL code examples in this chapter are deliberately dense (with reduced indentation and nesting), but the bold keywords help you to see the code structure. Most of the time, of course, we do not have the luxury of bold fonts (or color) to highlight code. In this case, you should add additional space, indentation, nesting, and comments.

Appendix A contains more detailed definitions and technical reference material.

TABLE 10.27 VHDL summary.

VHDL feature	Example	Book	93LRM
Comments	`-- this is a comment`	10.3	13.8

Literals (fixed-value items)	`12  1.0E6  '1'  "110"  'Z'`
	`2#1111_1111#    "Hello world"`
	`STRING'("110")`
		10.4	13.4

Identifiers (case-insensitive, start with letter)	`a_good_name    Same    same` `2_Bad    bad_    _bad    very__bad`	10.4	13.3
Several basic units of code	**`entity   architecture   configuration`**	10.5	1.1-1.3
Connections made through ports	**`port (signal in`** `i : BIT;` **`out`** `o : BIT);`	10.7	4.3
Default expression	**`port`** `(i : BIT := '1');` `-- i='1' if left open`	10.7	4.3
No built-in logic-value system. BIT and BIT_VECTOR (STD).	**`type`** `BIT` **`is`** `('0', '1'); -- predefined` **`signal`** `myArray: BIT_VECTOR (7` **`downto`** `0);`	10.8	14.2
Arrays	`myArray(1` **`downto`** `0) <= ('0', '1');`	10.8	3.2.1
Two basic types of logic signals	a `signal` corresponds to a real wire a `variable` is a memory location in RAM	10.9	4.3.1.2 4.3.1.3
Types and explicit initial/default value	**`signal`** `ONE : BIT := '1' ;`	10.9	4.3.2
Implicit initial/default value	`BIT'LEFT = '0'`	10.9	4.3.2
Predefined attributes	`clk'EVENT, clk'STABLE`	10.9.4	14.1
Sequential statements inside processes model things that happen one after another and repeat	**`process begin`** **`wait until`** `alarm = ring;` `eat; work; sleep;` **`end process;`**	10.10	8
Timing with wait statement	**`wait for`** `1 ns; -- not wait 1 ns` **`wait on`** `light` **`until`** `light = green;`	10.10.1	8.1
Update to signals occurs at the end of a simulation cycle	`signal <= 1; -- delta time delay` `signal <= variable1` **`after`** `2 ns;`	10.10.3	8.3
Update to variables is immediate	`variable := 1; -- immediate update`	10.10.3	8.4
Processes and concurrent statements model things that happen at the same time	**`process begin`** `rain ;` **`end process;`** **`process begin`** `sing ;` **`end process;`** **`process begin`** `dance;` **`end process;`**	10.13	9.2

IEEE Std_Logic_1164
(defines logic operators on 1164
types)
	`STD_ULOGIC, STD_LOGIC, STD_ULOGIC_VECTOR,` and
	`STD_LOGIC_VECTOR`
	`type` `STD_ULOGIC` **`is`**
	`('U','X','0','1','Z','W','L','H','-');`
		10.6	—

IEEE Numeric_Bit and Numeric_Std
(defines arithmetic operators on BIT
and 1164 types)
	`UNSIGNED` and `SIGNED`
	`X <= "10" * "01"`
	`-- OK with numeric pkgs.`
		10.12	—

TABLE 10.28   VHDL definitions.

Structure	Page	BNF	Structure	Page	BNF
alias declaration	418	10.21	next statement	429	10.32
architecture body	394	10.8	null statement	430	10.35
assertion statement	423	10.25	package declaration	398	10.11
attribute declaration	418	10.22	port interface declaration	406	10.13
block statement	438	10.37	port interface list	406	10.12
case statement	428	10.30	primary unit	393	10.5
component declaration	395	10.9	procedure call statement	427	10.28
component instantiation	444	10.42	process statement	440	10.38
concurrent statement	437	10.36	return statement	430	10.34
conditional signal assignment	442	10.40	secondary unit	393	10.6
configuration declaration	396	10.10	selected signal assignment	442	10.39
constant declaration	414	10.16	sequential statement	419	10.23
declaration	413	10.15	signal assignment statement	424	10.27
design file	393	10.4	signal declaration	414	10.17
entity declaration	394	10.7	special character	391	10.2
exit statement	430	10.33	subprogram body	416	10.20
generate statement	444	10.43	subprogram declaration	415	10.19
graphic character	391	10.1	type declaration	411	10.14
identifier	392	10.3	variable assignment statement	424	10.26
if statement	427	10.29	variable declaration	415	10.18
loop statement	429	10.31	wait statement	421	10.24

10.18 Problems

*=Difficult **=Very difficult ***=Extremely difficult

10.1 (Hello World, 10 min.) Set up a new, empty, directory (use `mkdir VHDL`, for example) to run your VHDL simulator (the exact details will depend on your computer and simulator). Copy the code below to a file called `hw_1.vhd` in your VHDL directory (leave out comments to save typing). *Hint:* Use the `vi` editor (`i` inserts text, `x` deletes text, `dd` deletes a line, `ESC :w` writes the file, `ESC :q` quits) or use `cat > hw_1.vhd` and type in the code (use `CTRL-D` to end typing) on a UNIX machine. Remember to save in 'Text Only' mode (Frame or MS Word) on an IBM PC or Apple Macintosh.

Analyze, elaborate, and simulate your model (include the output in your answer). Comment on how easy or hard it was to follow the instructions to use the software and suggest improvements.

```
entity HW_1 is end; architecture Behave of HW_1 is
constant M : STRING := "hello, world"; signal Ch : CHARACTER := ' ';
begin process begin
    for i in M'RANGE loop Ch <= M(i); wait for 1 ns; end loop; wait;
end process; end;
```

10.2 (Running a VHDL simulation, 20 min.) Copy the example from Section 10.1 into a file called `Counter1.vhd` in your VHDL directory (leave out the comments to save typing). Complete the compile (analyze), elaborate (build), and execute (initialize and simulate) or other equivalent steps for your simulator. After each step list the contents of your directory VHDL and any subdirectories and files that are created (use `ls -alR` on a UNIX system).

10.3 (Simulator commands, 10 min.) Make a "cheat sheet" for your simulator, listing the commands that can be used to control simulation.

10.4 (BNF addresses, 10 min.) Create a BNF description of a name including: optional title (Prof., Dr., Mrs., Mr., Miss, or Ms.), optional first name and middle initials (allow up to two), and last name (including unusual hyphenated and foreign names, such as Miss A-S. de La Salle, and Prof. John T. P. McTavish-fFiennes). The lowest level constructs are `letter ::= a-z`, `'.'` (period) and `'-'` (hyphen). Add BNF productions for a postal address in the form: company name, mail stop, street address, address lines (1 to 4), and country.

10.5 (BNF e-mail, 10 min.) Create a BNF description of a valid internet e-mail address in terms of letters, `'@'`, `'.'`, `'gov'`, `'com'`, `'org'`, and `'edu'`. Create a state diagram that "parses" an e-mail address for validity.

10.6 (BNF equivalence) Are the following BNF productions exactly equivalent? If they are not, produce a counterexample that shows a difference.

```
term ::= factor { multiplying_operator factor }
term ::= factor | term multiplying_operator factor
```

10.7 (Environment, 20 min.) Write a simple VHDL model to check and demonstrate that you can get to the IEEE library and have the environment variables, library statements, and such correctly set up for your simulator.

10.8 (Work, 20 min.) Write simple VHDL models to demonstrate that you can retrieve and use previously analyzed design units from the library `work` and that you can also remove design units from `work`. Explain how your models prove that access to `work` is functioning correctly.

10.9 (Packages, 60 min.) Write a simple package (use filename `PackH.vhd`) and package body (filename `PackB.vhd`). Demonstrate that you can store your package (call it `MyPackage`) in the library `work`. Then store, move, or rename (the details will depend on your software) your package to a library called `MyLibrary` in a directory called `MyDir`, and use its contents with a library clause (`library MyLibrary`) and a use clause (`use MyLibrary.MyPackage.all`) in a testbench called `PackTest` (filename `PackT.vhd`) in another directory `MyWork`. You may or may not be amazed at how complicated this can be and how poorly most software companies document this process.

10.10 (***IEEE Std 1164, 60 min.) Prior to VHDL-93 the `xnor` function was not available, and therefore older versions of the `std_logic_1164` library did not provide the `xnor` function for `STD_LOGIC` types either (it was actually included but commented out). Write a simple model that checks to see if you have the newer version of `std_logic_1164`. Can you do this without crashing the simulator?

You are an engineer on a very large project and find that your design fails to compile because your design must use the `xnor` function and the library setup on your company's system still points to the old IEEE `std_logic_1164` library, even though the new library was installed. You are apparently the first person to realize the problem. Your company has a policy that any time a library is changed all design units that use that library must be rebuilt from source. This might require days or weeks of work. Explain in detail, using code, the alternative solutions. What will you recommend to your manager?

10.11 (**VHDL-93 test, 20 min.) Write a simple test to check if your simulator is a VHDL-87 or VHDL-93 environment—without crashing the simulator.

10.12 (Declarations, 10 min.) Analyze the following changes to the code in Section 10.8 and include the simulator output in your answers:

a. Uncomment the declarations for `Bad100` and `Bad4` in `Declaration_1`.

b. Add the following to `Constant_2`:

```
signal wacky : wackytype (31 downto 0); -- wacky
```

c. Remove the library and use clause in `Constant_2`.

10.13 (`STRING` type, 10 min.) Replace the `write` statement that prints the string `" count="` in `Text(Behave)` in Section 10.6.3 with the following, compile it, and explain the result:

```
write(L, " count=" ); -- No type qualification.
```

10.14 (Sequential statements, 10 min.) Uncomment the following line in Wait_1(Behave) in Section 10.10, analyze the code, and explain the result:

```
wait on x(1 to v); -- v is a variable.
```

10.15 (VHDL logical operators, 10 min.)

a. Explain the problem with the following VHDL statement:

```
Z <= A nand B nand C;
```

b. Explain why this problem does not occur with this statement:

```
Z <= A and B and C;
```

c. What can you say about the logical operators: and, or, nand, nor, xnor, xor?

d. Is the following code legal?

```
Z <= A and B or C;
```

10.16 (*Initialization, 45 min.) Consider the following code:

```
entity DFF_Plain is port (Clk, D : in BIT; Q : out BIT); end;
architecture Bad of DFF_Plain is begin process (Clk) begin
   if Clk = '0' and Clk'EVENT then Q <= D after 1 ns; end if;
end process; end;
```

a. Analyze and simulate this model using a testbench.

b. Rewrite architecture Bad using an equivalent process including a wait statement. Simulate this equivalent model and confirm the behaviors are identical.

c. What is the behavior of the output Q during initial execution of the process?

d. Why does this happen?

e. Why does this not happen with the following code:

```
architecture Good of DFF_Plain is
begin process begin wait until Clk = '0'; Q <= D after 1 ns;
end process; end;
```

10.17 (Initial and default values, 20 min.) Use code examples to explain the difference between: default expression, default value, implicit default value, initial value, initial value expression, and default initial value.

10.18 (Enumeration types, 20 min.) Explain the analysis results for the following:

```
type MVL4 is ('X', '0', '1', 'Z'); signal test : MVL4;
process begin
   test <= 1; test <= Z; test <= z; test <= '1'; test <= 'Z';
end process;
```

Alter the type declaration to the following, analyze your code again, and comment:

```
type Mixed4 is (X , '0', '1', Z);
```

10.19 (Type declarations, 10 min.) Correct these declarations:

```
type BadArray is array (0 to 7) of BIT_VECTOR;
type Byte is array (NATURAL range 7 downto 0) of BIT;
subtype BadNibble is Byte(3 downto 0);
type BadByte is array (range 7 downto 0) of BIT;
```

10.20 (Procedure parameters, 10 min.) Analyze the following package; explain and correct the error. Finally, build a testbench to check your solution.

```
package And_Pkg_Bad is procedure V_And(a, b : BIT; c: out BIT); end;
package body And_Pkg_Bad is
procedure V_And(a,b : BIT;c : out BIT) is begin c <= a and b;end;
end And_Pkg_Bad;
```

10.21 (Type checking, 20 min.) Test the following code and explain the results:

```
type T is INTEGER range 0 to 32; variable a: T;
a := (16 + 17) - 12; a := 16 - 12 + 17; a := 16 + (17 - 12);
```

10.22 (Debugging VHDL code, 30 min.) Find and correct the errors in the following code. Create a testbench for your code to check that it works correctly.

```
entity UpDownCount_Bad is
port(clock, reset, up: STD_LOGIC; D: STD_LOGIC_VECTOR (7 to 0));
end UpDownCount_Bad;

architecture Behave of UpDownCount_Bad is
begin process (clock, reset, up); begin
if (reset = '0') then D <= '0000000';
elseif (rising_edge(clock)) then
if (up = 1) D <= D+1; else D <= D-1; end if;
end if; end process; end Behave;
```

10.23 (Subprograms, 20 min.) Write and test subprograms for these declarations:

```
function Is_X_Zero (signal X : in BIT) return BIT;

procedure Is_A_Eq_B (signal A, B : BIT; signal Y : out BIT);
```

10.24 (Simulator error messages, 10 min.) Analyze and attempt to simulate `Arithmetic_2(Behave)` from Section 10.12 and compare the error message you receive with that from the MTI simulator (not all simulators are as informative). There are no standards for error messages.

10.25 (Exhaustive property of case statement, 30 min.) Write and simulate a testbench for the state machine of Table 10.8 and include your results. Is every state transition tested by your program and is every transition covered by an assignment statement in the code? (*Hint:* Think very carefully.) Repeat this exercise for the state machine in Section 10.10.6.

10.26 (Default values for inputs, 20 min.) Replace the interface declaration for entity `Half_Adder` in Section 10.5 with the following (to remove the default values):

```
port (X, Y: in BIT ; Sum, Cout: out BIT);
```

Attempt to compile, elaborate, and simulate configuration `Simplest` (the other entities needed, `AndGate` and `XorGate`, must already be in `work` or in the same file). You should get an error at some stage (different systems find this error at different points—just because an entity compiles, that does not mean it is error-free).

The LRM says "... A port of mode in may be unconnected ...only if its declaration includes a **default expression**..." [VHDL 93LRM1.1.1.2].

We face a dilemma here. If we do not drive inputs with test signals and leave an input port unconnected, we can compile the model (since it is syntactically correct) but the model is not semantically correct. On the other hand, if we give the inputs default values, we might accidentally forget to make a connection and not notice.

10.27 (Adder generation, 10 min.) Draw the schematic for `Adder_1(Structure)` of Section 10.13.7, labeling each instance with the VHDL instance name.

10.28 (Generate statement, 20 min.) Draw a schematic corresponding to the following code (label the cells with their instance names):

```
B1: block begin L1 : C port map (T, B, A(0), B(0)) ;
L2: for i in 1 to 3 generate L3 : for j in 1 to 3 generate
L4: if i+j > 4 generate L5: C port map (A(i—1), B(j—1), A(i), B(j)) ;
end generate; end generate; end generate;
L6: for i in 1 to 3 generate L7: for j in 1 to 3 generate
L8: if i+j < 4 generate L9: C port map (A(i+1), B(j+1), A(i), B(j)) ;
end generate; end generate; end generate;
end block B1;
```

Rewrite the code without `generate` statements. How would you prove that your code really is exactly equivalent to the original?

10.29 (Case statement, 20 min.) Create a package (`my_equal`) that overloads the equality operator so that `'X'='0'` and `'X'='1'` are both `TRUE`. Test your package. Simulate the following design unit and explain the result.

```
entity Case_1 is end; architecture Behave of Case_1 is
signal r : BIT; use work.my_equal.all;
begin process variable twobit:STD_LOGIC_VECTOR(1 to 2); begin
  twobit := "X0";
  case twobit is
    when "10" => r <= '1';
    when "00" => r <= '1';
    when others => r <= '0';
  end case; wait;
end process; end;
```

10.30 (State machine) Create a testbench for the state machine of Section 10.2.5.

10.31 (Mealy state machine, 60 min.) Rewrite the state machine of Section 10.2.5 as a Mealy state machine (the outputs depend on the inputs and on the current state).

10.32 (Gate-level D flip-flop, 30 min.) Draw the schematic for the following D flip-flop model. Create a testbench (check for correct operation with combinations of Clear, Preset, Clock, and Data). Have you covered all possible modes of operation? Justify your answer of yes or no.

```
architecture RTL of DFF_To_Test is
signal A, B, C, D, QI, QBarI : BIT; begin
A <= not (Preset and D and B) after 1 ns;
B <= not (A and Clear and Clock) after 1 ns;
C <= not (B and Clock and D) after 1 ns;
D <= not (C and Clear and Data) after 1 ns;
QI <= not (Preset and B and QBarI) after 1 ns;
QBarI <= not (QI and Clear and C) after 1 ns;
Q <= QI; QBar <= QBarI;
end;
```

10.33 (Flip-flop model, 20 min.) Add an asynchronous active-low preset to the D flip-flop model of Table 10.3. Generate a testbench that includes interaction of the preset and clear inputs. What issue do you face and how did you solve it?

10.34 (Register, 45 min.) Design a testbench for the register of Table 10.4. Adapt the 8-bit register design to a 4-bit version with the following interface declaration:

```
entity Reg4 is port (D : in  STD_LOGIC_VECTOR(7 downto 0);
Clk,Pre,Clr : in  STD_LOGIC;Q,QB : out STD_LOGIC_VECTOR(7 downto 0));
end Reg8;
```

Create a testbench for your 4-bit register with the following component declaration:

```
component DFF
port(Preset,Clear,Clock,Data:STD_LOGIC;Q,QBar:out STD_LOGIC_VECTOR);
end component;
```

10.35 (*Conversion functions, 30 min.) Write a conversion function from NATURAL to STD_LOGIC_VECTOR using the following declaration:

```
function Convert (N, L: NATURAL) return STD_LOGIC_VECTOR;
-- N is NATURAL, L is length of STD_LOGIC_VECTOR
```

Write a similar conversion function from STD_LOGIC_VECTOR to NATURAL:

```
function Convert (B: STD_LOGIC_VECTOR) return NATURAL;
```

Create a testbench to test your functions by including them in a package.

10.36 (Clock procedure, 20 min.) Design a clock procedure for a two-phase clock (C1, C2) with variable high times (HT1, HT2) and low times (LT1, LT2) and the following interface. Include your procedure in a package and write a model to test it.

```
procedure Clock (C1, C2 : out STD_LOGIC; HT1, HT2, LT1, LT2 : TIME);
```

10.37 (Random number, 20 min.) Design a testbench for the following procedure:

```
procedure uniform (seed : inout INTEGER range 0 to 15) is
   variable x : INTEGER;
   begin x := (seed*11) + 7; seed := x mod 16;
end uniform;
```

10.38 (Full-adder, 30 min.) Design and test a behavioral model of a full adder with the following interface:

```
entity FA is port (X, Y, Cin : STD_LOGIC; Cout, Sum : out STD_LOGIC);
end;
```

Repeat the exercise for inputs and outputs of type UNSIGNED.

10.39 (8-bit adder testbench, 60 min.) Write out the code corresponding to the generate statements of Adder_1 (Structure) in Section 10.13.7. Write a testbench to check your adder. What problems do you encounter? How thorough do you believe your tests are?

10.40 (Shift-register testbench, 60 min.) Design a testbench for the shift register of Table 10.4. Convert this model to use STD_LOGIC types with the following interface:

```
entity ShiftN is
port (CLK, CLR, LD, SH, DIR : STD_LOGIC;
   D : STD_LOGIC_VECTOR; Q : out STD_LOGIC_VECTOR);
end;
```

10.41 (Multiplier, 60 min.) Design and test a multiplier with the following interface:

```
entity Mult8 is
port (A, B : STD_LOGIC_VECTOR(3 downto 0);
Start, CLK, Reset : in STD_LOGIC;
Result : out STD_LOGIC_VECTOR(7 downto 0); Done : out BIT);
end;
```

a. Create testbench code to check your model.

b. Catalog each compile step with the syntax errors as you debug your code.

c. Include a listing of the first code you write together with the final version.

An interesting class project is to collect statistics from other students working on this problem and create a table showing the types and frequency of syntax errors made with each compile step, and the number of compile steps required. Does this information suggest ways that you could improve the compiler, or suggest a new type of tool to use when writing VHDL?

10.42 (Port maps, 5 min.) What is wrong with this VHDL statement?

```
U1 : nand2 port map (a <= set, b <= qb, c <= q);
```

10.43 (DRIVING_VALUE, 15 min.) Use the VHDL-93 attribute Clock'DRIVING_VALUE to rewrite the following clock generator model without using a temporary variable.

```
entity ClockGen_2 is port (Clock : out BIT); end;
architecture Behave of ClockGen_2 is
begin process variable Temp : BIT := '1'; begin
  Temp := not Temp ; Clock <= Temp after 10 ns; wait for 10 ns;
  if (now > 100 ns) then wait; end if; end process;
end;
```

10.44 (Records, 15 min.) Write an architecture (based on the following skeleton) that uses the record structure shown:

```
entity Test_Record_1 is end; architecture Behave of Test_Record_1 is
begin process type Coordinate is record X, Y : INTEGER; end record;
-- a record declaration for an attribute declaration:
attribute Location:Coordinate; -- an attribute declaration
begin wait; end process; end Behave;
```

10.45 (**Communication between processes, 30 min.) Explain and correct the problem with the following skeleton code:

```
variable v1 : INTEGER := 1; process begin v1 := v1+3; wait; end process;
process variable v2 : INTEGER := 2; begin v2 := v1  ; wait; end process;
```

10.46 (*Resolution, 30 min.) Explain and correct the problems with the following:

```
entity R_Bad_1 is port (i : in BIT; o out BIT); end;
architecture Behave of R_Bad_1 is
begin o <= not i after 1 ns; o <= i after 2 ns; end;
```

10.47 (*Inputs, 30 min.) Analyze the following and explain the result:

```
entity And2 is port (A1, A2: in BIT; ZN: out BIT); end;
architecture Simple of And2 is begin ZN <= A1 and A2; end;

entity Input_Bad_1 is end; architecture Netlist of Input_Bad_1 is
component And2 port (A1, A2 : in BIT; ZN : out BIT); end component;
signal X, Z : BIT begin G1 : And2 port map (X, X, Z); end;
```

10.48 (Association, 15 min.) Analyze the following and explain the problem:

```
entity And2 is port (A1, A2 : in BIT; ZN : out BIT); end;
architecture Simple of And2 is begin ZN <= A1 and A2; end;

entity Assoc_Bad_1 is port (signal X, Y : in BIT; Z : out BIT); end;
architecture Netlist of Assoc_Bad_1 is
component And2 port (A1, A2 : in BIT; ZN : out BIT); end component;
```

```
begin
G1: And2 port map (X, Y, Z);
G2: And2 port map (A2 => Y, ZN => Z, A1 => X);
G3: And2 port map (X, ZN => Z, A2 => Y);
end;
```

10.49 (Modes, 30 min.) Analyze and explain the errors in the following:

```
entity And2 is port (A1, A2 : in BIT; ZN : out BIT); end;
architecture Simple of And2 is begin ZN <= A1 and A2; end;

entity Mode_Bad_1 is port (X : in BIT; Y : out BIT; Z : inout BIT); end;
architecture Netlist of Mode_Bad_1 is
component And2 port (A1, A2 : in BIT; ZN : out BIT); end component;
begin G1 : And2 port map (X, Y, Z); end;

entity Mode_Bad_2 is port (X : in BIT; Y : out BIT; Z : inout BIT); end;
architecture Netlist of Mode_Bad_1 is
component And2 port (A1, A2 : in BIT; ZN : inout BIT); end component;
begin G1 : And2 port map (X, Y, Z); end;
```

10.50 (*Mode association, 60 min.) Analyze and explain the errors in the following code. The number of errors, types of error, and the information in the error messages given by different simulators vary tremendously in this area.

```
entity Allmode is port
(I : in BIT; O : out BIT; IO : inout BIT; B : buffer BIT);
end;
architecture Simple of Allmode is begin O<=I; IO<=I; B<=I; end;

entity Mode_1 is port
(I : in BIT; O : out BIT; IO : inout BIT; B : buffer BIT);
end;
architecture Netlist of Mode_1 is
component Allmode port
(I : in BIT; O : out BIT; IO : inout BIT; B : buffer BIT); end
component;
begin
G1:Allmode port map (I , O , IO, B );
G2:Allmode port map (O , IO, B , I );
G3:Allmode port map (IO, B , I , O );
G4:Allmode port map (B , I , O , IO);
end;
```

10.51 (**Declarations, 60 min.) Write a tutorial (approximately two pages of text, five pages with code) with examples explaining the difference between: a component declaration, a component configuration, a configuration declaration, a configuration specification, and a block configuration.

10.52 (**Guards and guarded signals, 60 min.) Write some simple models to illustrate the use of guards, guarded signals, and the disconnect statement. Include an experiment that shows and explains the use of the implicit signal GUARD in assignment statements.

10.53 (**Std_logic_1164, 120 min.) Write a short (two pages of text) tutorial, with (tested) code examples, explaining the std_logic_1164 types, their default values, the difference between the 'ulogic' and 'logic' types, and their vector forms. Include an example that shows and explains the problem of connecting a std_logic_vector to a std_ulogic_vector.

10.54 (Data swap, 20 min.) Consider the following code:

```
library ieee; use ieee.std_logic_1164.all;
package config is
type type1 is record
f1 : std_logic_vector(31 downto 0); f2 : std_logic_vector(3 downto 0);
end record;
type type2 is record
f1 : std_logic_vector(31 downto 0); f2 : std_logic_vector(3 downto 0);
end record;
end config;
library ieee; use ieee.STD_LOGIC_1164.all; use work.config.all;
entity Swap_1 is
port (Data1 : type1; Data2 : type2; sel : STD_LOGIC;
Data1Swap : out type1; Data2Swap : out type2); end Swap_1;

architecture Behave of Swap_1 is begin
Swap: process (Data1, Data2, sel) begin case sel is
when '0' => Data1Swap <= Data1; Data2Swap <= Data2;
when others => Data1Swap <= Data2; Data2Swap <= Data1;
end case; end process Swap; end Behave;
```

Compile this code. What is the problem? Suggest a fix. Now write a testbench and test your code. Have you considered all possibilities?

10.55 (***RTL, 30 min.) "**RTL** stands for **register-transfer level**. ...when referencing VHDL, the term means that the description includes only concurrent signal assignment statements and possibly block statements. In particular, VHDL data flow descriptions explicitly do not contain either process statements (which describe behavior) or component instantiation statements (which describe structure)" (Dr. VHDL from VHDL International).

    a. With your knowledge of process statements and components, comment on Dr. VHDL's explanation.

    b. In less than 100 words offer your own definition of the difference between RTL, data flow, behavioral, and structural models.

10.56 (*Operators mod and rem, 20 min.) Confirm and explain the following:

```
i1 := (-12)  rem  7;          -- i1 = -5
i2 := 12     rem  (-7);       -- i2 =  5
i3 := (12)   rem  (-7);       -- i3 = -5
i4 := 12     mod  7;          -- i4 =  5
i5 := (-12)  mod  7;          -- i5 =  2
i6 := 12     mod  (-7);       -- i6 = -2
i7 := (12)   mod  (-7);       -- i7 = -5
```

Evaluate –5 `rem` 2 and explain the result.

10.57 (***Event and stable, 60 min.) Investigate the differences between `clk'EVENT` and `clk'STABLE`. Write a minitutorial (in the form of a "cheat sheet") with examples showing the differences and potential dangers of using `clk'STABLE`.

10.58 (PREP benchmark #2, 60 min.) The following code models a benchmark circuit used by **PREP** to measure the capacity of FPGAs. Rewrite the concurrent signal assignment statements (labeled `mux` and `comparator`) as equivalent processes. Draw a datapath schematic corresponding to `PREP2(Behave_1)`. Write a testbench for the model. Finally (for extra credit) rewrite the model and testbench to use `STD_LOGIC` instead of `BIT` types.

```
library ieee; use ieee.STD_LOGIC_1164.all;
use ieee.NUMERIC_BIT.all; use ieee.NUMERIC_STD.all;
entity PREP2 is
port(CLK,Reset,Sel,Ldli,Ldhi : BIT; D1,D2 : STD_LOGIC_VECTOR(7 downto 0);
   DQ:out STD_LOGIC_VECTOR(7 downto 0));
end;

architecture Behave_1 of PREP2 is
signal EQ : BIT; signal y,lo,hi,Q_i : STD_LOGIC_VECTOR(7 downto 0);
begin
outputDriver: Q <= Q_i;
mux: with Sel select y <= hi when '0', D1 when '1';
comparator: EQ <= '1' when Q_i = lo else '0';
   register: process(Reset, CLK) begin
      if Reset = '1' then hi <= "00000000"; lo <= "00000000";
      elsif CLK = '1' and CLK'EVENT then
         if Ldhi='1' then hi<=D2;end if;if Ldlo='1' then lo<=D2;end if;
      end if;
   end process register;
   counter: process(Reset, CLK) begin
      if Reset = '1' then Q_i <= "00000000";
      elsif CLK = '1' and CLK'EVENT then
         if EQ = '1' then Q_i <= y;
         elsif EQ = '0' then Q_i   <= Q_i + "00000001";
         end if;
      end if;
   end process counter;
end;
```

10.59 (PREP #3, state machine) Draw the state diagram for the following PREP benchmark (see Problem 10.58). Is this a Mealy or Moore machine? Write a testbench and test this code.

```
library ieee; use ieee.STD_LOGIC_1164.all;
entity prep3_1 is port(Clk, Reset: STD_LOGIC;
   I : STD_LOGIC_VECTOR(7 downto 0); O : out STD_LOGIC_VECTOR(7 downto 0));
end prep3_1;
architecture Behave of prep3_1 is
```

```
      type STATE_TYPE is (sX,s0,sa,sb,sc,sd,se,sf,sg);
      signal state : STATE_TYPE; signal Oi : STD_LOGIC_VECTOR(7 downto 0);
begin
  O <= Oi;
  process (Reset, Clk) begin
    if (Reset = '1') then state <= s0; Oi <= (others => '0');
    elsif rising_edge(Clk) then
      case state is
        when s0 =>
          if (I = X"3c") then state <= sa; Oi <= X"82";
          else state <= s0; Oi <= (others => '0');
          end if;
        when sa =>
          if (I = X"2A") then state <= sc; Oi <= X"40";
          elsif (I = X"1F") then state <= sb; Oi <= X"20";
          else state <= sa; Oi <= X"04";
          end if;
        when sb =>
          if (I = X"AA") then state <= se; Oi <= X"11";
          else state <= sf; Oi <= X"30";
          end if;
        when sc => state <= sd; Oi <= X"08";
        when sd => state <= sg; Oi <= X"80";
        when se => state <= s0; Oi <= X"40";
        when sf => state <= sg; Oi <= X"02";
        when sg => state <= s0; Oi <= X"01";
        when others => state <= sX; Oi <= (others => 'X');
      end case;
    end if;
  end process;
end;
```

10.60 (Edge detection, 30 min) Explain the construction of the IEEE 1164 function to detect the rising edge of a signal, `rising_edge(s)`. List all the changes in signal s that correspond to a rising edge.

```
function rising_edge (signal s : STD_ULOGIC) return BOOLEAN is
  begin return
  (s'EVENT and (To_X01(s) = '1') and (To_X01(s'LAST_VALUE) = '0'));
end;
```

10.61 (*Real, 10 min.) Determine the smallest real in your VHDL environment.

10.62 (*Stop, 30 min.) How many ways are there to stop a VHDL simulator?

10.63 (*Arithmetic package, 60 min.) Write a function for an arithmetic package to subtract two's complement numbers. Create a test bench to check your function. Your declarations in the package header should look like this:

```
type TC is array (INTEGER range <>) of STD_LOGIC;
function "-"(L : TC; R : TC) return TC;
```

10.64 (***Reading documentation, hours) There are a few gray areas in the interpretation of the VHDL-87 LRM some of which were clarified in the VHDL-93 revision. One VHDL system has a "**compatibility mode**" that allows alternative interpretations. For each of the following "issues" taken from the actual tool documentation try to interpret what was meant, determine the interpretation taken by your own software, and then rewrite the explanation clearly using examples.

a. * "Unassociated variable and signal parameters. Compatibility mode allows variable and signal parameters to subprograms to be unassociated if they have a default value. Otherwise, an error is generated."

Example answer: Consider the following code:

```
package Util_2 is
procedure C(signal Clk : out BIT; signal P : TIME := 10 ns);
end Util_2;
package body Util_2 is
procedure C(signal Clk : out BIT; signal P : TIME := 10 ns) is
begin loop Clk <= '1' after P/2, '0' after P;
wait for P; end loop; end; end Util_2;
entity Test_Compatibility_1 is end; use work.Util_2.all;
architecture Behave of Test_Compatibility_1 is
signal v,w,x,y,z : BIT; signal s : TIME := 5 ns;
begin process variable v : TIME := 5 ns; begin
C(v, s);          -- parameter s is OK since P is declared as signal
-- C(w, v);       -- would be OK if P is declared as variable instead
-- C(x, 5 ns);    -- would be OK if P is declared as constant instead
-- C(y);          -- unassociated, an error if P is signal or variable
-- C(z,open);     -- open, an error if P is signal or variable
end process; end;
```

The Compass Scout simulator (which does not have a compatibility mode) generates an error during analysis if a signal or variable subprogram parameter is open or unassociated (a constant subprogram parameter may be unassociated or open).

b. * "Allow others in an aggregate within a record aggregate. The LRM [7.3.2.2] defines nine situations where others may appear in an aggregate. In compatibility mode, a tenth case is added. In this case, others is allowed in an aggregate that appears as an element association in a record element."

c. * "BIT'('1') parsed as BIT ' ('1'). The tick (') character is being used twice in this example. In the first case as an attribute indicator, in the second case, to form a character literal. Without the compatibility option, the analyzer adopts a strict interpretation of the LRM, and without white space around the first tick, the fragment is parsed as BIT '('1'), that is, the left parenthesis ('(') is the character literal."

d. ** "Generate statement declarative region. Generate statements form their own declarative region. In compatibility mode, configuration specifications will apply to items being instantiated within a generate statement."

e. ** "Allow type conversion functions on open parameters. If a parameter is specified as open, it indicates a parameter without an explicit association. In such cases, the presence of a type conversion function is meaningless. Compatibility mode allows the type conversion functions."

f. *** "Entity class flexibility. Section [3.1.2] of the LRM defines the process of creating a new integer type. The type name given is actually assigned to a subtype name, related to an anonymous base type. This implies that the entity class used during an attribute specification [LRM 5.1] should indicate subtype, rather than type. Because the supplied declaration was type rather than subtype, compatibility mode allows type."

g. *** "Allowing declarations beyond an all/others specification. Section [5.1] of the LRM states that the first occurrence of the reserved word `all` or `others` in an attribute specification terminates the declaration of the related entity class. The LRM declares that the entity/architecture and package/package body library units form single declaration regions [LRM 10.1] that are the concatenation of the two individual library declarative regions. For example, if a signal attribute specification with `all` or `others` was specified in the entity, it would be impossible to declare a signal in the architecture. In compatibility mode, this LRM limitation is removed."

h. *** "User-defined attributes on overloaded functions. In compatibility mode, user-defined attributes are allowed to be associated with overloaded functions. Note: Even in compatibility mode, there is no way to retrieve the different attributes."

10.65 (*1076 interpretations, 30 min.) In a DAC paper, the author writes: 'It was experienced that (company R) might have interpreted IEEE 1076 differently than (company S) did, e.g. concatenations (&) are not allowed in "case selector" expressions for (company S).' Can you use concatenation in your VHDL tool for either the `expression` or `choices` for a `case` statement?

10.66 (**Interface declarations, 15 min.) Analyze the following and comment:

```
entity Interface_1 is
  generic (I : INTEGER; J : INTEGER := I; K, L : INTEGER);
  port (A : BIT_VECTOR; B : BIT_VECTOR(A'RANGE); C : BIT_VECTOR (K to L));
  procedure X(P, Q : INTEGER; R : INTEGER range P to Q);
  procedure Y(S : INTEGER range K to L);
end Interface_1;
```

10.67 (**Wait statement, 10 min.) Construct the sensitivity set and thus the sensitivity list for the following `wait` statement (that is, rewrite the `wait` statement in the form `wait on sensitivity_list until condition`).

```
entity Complex_Wait is end;                              --1
architecture Behave of Complex_Wait is                   --2
  type A is array (1 to 5) of BOOLEAN;                   --3
```

```
function F (P : BOOLEAN) return BOOLEAN;                        --4
signal S : A; signal i, j : INTEGER range 1 to 5;              --5
begin process begin                                            --6
  wait until F(S(3)) and (S(i) or S(j));                       --7
end process;                                                   --8
end;                                                           --9
```

10.68 (**Shared variables, 20 min.) Investigate the following code and comment:

```
architecture Behave of Shared_1 is
subtype S is INTEGER range 0 to 1; shared variable C : S := 0; begin
process begin C := C + 1; wait; end process;
process begin C := C - 1; wait; end process;
end;
```

10.69 (Undocumented code and ranges, 20 min.) Explain the purpose of the following function (part of a package from a well-known synthesis company) with a parameter of type SIGNED. Write a testbench to check your explanation. Investigate what happens when you call this function with a string-literal argument, for example with the statement X <= IM("11100"). What is the problem and why does it happen? Rewrite the code, including documentation, to avoid this problem.

```
type SIGNED is array (NATURAL range <> ) of BIT;

function IM (L : SIGNED) return INTEGER is variable M : INTEGER;
begin M := L'RIGHT-1;
  for i in L'LEFT-1 downto L'RIGHT loop
    if (L(i) = (not L(L'LEFT))) then M := i; exit; end if;
  end loop; return M;
end;
```

10.70 (Timing parameters, 20 min.) Write a model and a testbench for a two-input AND gate with separate rising (tpLH) and falling (tpHL) delays using the following interface:

```
entity And_Process is
generic (tpLH, tpHL : TIME); port (a, b : BIT; z : out BIT) end;
```

10.71 (Passive code in entities, 30 min.) Write a procedure (CheckTiming, part of a package Timing_Pkg) to check that two timing parameters (tPLH and tPHL) are both greater than zero. Include this procedure in a two-input AND gate model (And_Process). Write a testbench to show your procedure and gate model both work. Rewrite the entity for And_Process to include the timing check as part of the entity declaration. You are allowed to include **passive code** (no assignments to signals and so on) directly in each entity. This avoids having to include the timing checks in each architecture.

10.72 (Buried code, 30 min.) Some companies bury instructions to the software within their packages. Here is an example of part of the arithmetic package from an imaginary company called SissyN:

```
function UN_plus(A, B : UN) return UN is                          --1
variable CRY : STD_ULOGIC; variable X,SUM : UN (A'LEFT downto 0);  --2
-- pragma map_to_operator ADD_UNS_OP                               --3
-- pragma type_function LEFT_UN_ARG                                --4
-- pragma return_port_name Z                                       --5
begin                                                             --6
-- sissyn synthesis_off                                            --7
if (A(A'LEFT) = 'X' or B(B'LEFT) = 'X') then SUM := (others => 'X'); --8
return(SUM);                                                       --9
end if;                                                          --10
-- sissyn synthesis_on                                           --11
CRY := '0'; X := B;                                              --12
for i in 0 to A'LEFT loop                                        --13
SUM(i) := A(i) xor X(i) xor carry;                              --14
CRY := (A(i) and X(i)) or (A(i) and CRY) or (CRY and X(i));     --15
end loop; return SUM;                                           --16
end;                                                            --17
```

Explain what this function does. Can you now hazard a guess at what each of the comments means? What are the repercussions of using comments in this fashion?

10.73 (*Deferred constants, 15 min.) "If the assignment symbol `':='` followed by an expression is not present in a constant declaration, then the declaration declares a **deferred constant**. Such a constant declaration may only appear in a package declaration. The corresponding full constant declaration, which defines the value of the constant, must appear in the body of the package" [VHDL 93LRM4.3.1.1].

```
package Constant is constant s1, s2 : BIT_VECTOR; end Constant;

package body Constant is
constant s0 : BIT_VECTOR := "00"; constant s1 : BIT_VECTOR := "01";
end Constant;
```

It is tempting to use deferred constants to hide information. However, there are problems with this approach. Analyze the following code, explain the results, and correct the problems:

```
entity Deferred_1 is end; architecture Behave of Deferred_1 is
use work.all; signal y,i1,i2 : INTEGER; signal sel : INTEGER range 0 to 1;
begin with sel select y <= i1 when s0, i2 when s1; end;
```

10.74 (***Viterbi code, days) Convert the Verilog model of the Viterbi decoder in Chapter 11 to VHDL. This problem is tedious without the help of some sort of **Verilog to VHDL conversion** process. There are two main approaches to this problem. The first uses a synthesis tool to read the behavioral Verilog and write structural VHDL (the Compass ASIC Synthesizer can do this, for example). The second approach uses conversion programs (Alternative System Concepts Inc. at `http://www.ascinc.com` is one source). Some of these companies allow you to e-mail code to them and they will automatically return a translated version.

10.75 (*Wait statement, 30 min.) Rewrite the code below using a single `wait` statement and write a testbench to prove that both approaches are exactly equivalent:

```
entity Wait_Exit is port (Clk : in BIT); end;
architecture Behave of Wait_Exit is
  begin process begin
    loop wait on Clk; exit when Clk = '1'; end loop;
  end process;
end;
```

10.76 (Expressions, 10 min.) Explain and correct the problems with the following:

```
variable b : BOOLEAN; b := "00" < "11";                         --1
variable bv8 : BIT_VECTOR (7 downto 0) := "1000_0000";          --2
```

10.77 (Combinational logic using `case` statement, 10 min.) A Verilog user suggests the following method to model combinational logic. What are the problems with this approach? Can you get it to work?

```
entity AndCase is port (a, b : BIT; y : out BIT); end;
architecture Behave of AndCase is begin process (a , b) begin
  case a & b is
    when '1'&'1' => y <= '1'; when others => y <= '0';
  end case;
end process; end;
```

10.78 (*Generics and back-annotation, 60 min.)

a. Construct design entities `And_3(Behave)`, a two-input AND gate, and `Xor_3(Behave)`, a two-input XOR gate. Include generic constants to model the propagation delay from each input to the output separately. Use the following entity declaration for `And_3`:

```
entity And_3 is port (I1, I2 : BIT; O : out BIT);
  generic (I1toO, I2toO : DELAY_LENGTH := 0.4 ns); end;
```

b. Create and test a package, `P_1`, that contains `And_3` and `Xor_3` as components.

c. Create and test a design entity `Half_Adder_3(Structure_3)` that uses `P_1`, with the following interface:

```
entity Half_Adder_3 is port (X, Y : BIT; Sum, Carry : out BIT); end;
```

d. Modify and test the architecture `Structure_3` for `Half_Adder_3` so that you can use the following configuration:

```
configuration Structure_3 of Half_Adder_3 is
for Structure_3
for L1 : XOR generic map (0.66 ns,0.69 ns); end for;
for L2 : AND generic map (0.5 ns, 0.6 ns) port map (I2 => HI); end for;
end for; end;
```

10.79 (SNUG'95, *60 min.) In 1995 John Cooley organized a contest between VHDL and Verilog for ASIC designers. The goal was to design the fastest 9-bit counter in under one hour using Synopsys synthesis tools and an LSI Logic vendor technology library. The VHDL interface is as follows:

```
library ieee; use ieee.std_logic_1164.all;
-- use ieee.std_logic_arith.all; -- substitute your package here
entity counter is port (
data_in    : in std_logic_vector(8 downto 0);
up         : in std_logic;
down       : in std_logic;
clock      : in std_logic;
count_out  : inout std_logic_vector(8 downto 0);
carry_out  : out std_logic;
borrow_out : out std_logic;
parity_out : out std_logic ); end counter;
architecture example of counter is begin
-- insert your design here
end example;
```

The counter is positive-edge triggered, counts up with up = '1' and down with down = '1'. The contestants had the advantage of a predefined testbench with a set of test vectors, you do not. Design a model for the counter and a testbench. How confident are you that you have thoroughly tested your model? (In the real contest none of the VHDL contestants managed to even complete a working design in under one hour. In addition, the VHDL experts that had designed the testbench omitted a test case for one of the design specifications.)

10.80 (*A test procedure, 45 min.) Write a procedure all (for a package test) that serially generates all possible input values for a signal spaced in time by a delay, dly. Use the following interface:

```
library ieee; use ieee.std_logic_1164.all; package test is
procedure all (signal SLV : out STD_LOGIC_VECTOR; dly : in TIME);
end package test ;
```

10.81 (Direct instantiation, 20 min.) Write an architecture for a full-adder, entity Full_Adder_2, that directly instantiates units And_2(Behave) and Xor_2(Behave). This is only possible in a VHDL-93 environment.

```
entity And_2 is port (i1, i2 : BIT; y : out BIT); end;
entity Xor_2 is port (i1, i2 : BIT; y : out BIT); end;
entity Full_Adder_2 is port (a, b, c : BIT ; sum, cout : out BIT); end;
```

10.82 (**Shift operators for 1164, 60 min.) Write a package body to implement the VHDL-93 shift operators, sll and srl, for the type STD_LOGIC_VECTOR. Use the following package header:

```
package 1164_shift is
function "sll"(x : STD_LOGIC_VECTOR; n : INTEGER)
   return STD_LOGIC_VECTOR;
```

```
function "srl"(x : STD_LOGIC_VECTOR; n : INTEGER)
  return STD_LOGIC_VECTOR;
end package 1164_shift;
```

10.83 (**VHDL `wait` statement, 60 min.) What is the problem with the following VHDL code? *Hint:* You may need to consult the VHDL LRM.

```
procedure p is begin wait on b; end;
process (a) is begin procedure p; end process;
```

10.84 (**Null range, 45 min.) A range such as 1 `to` -1 or 0 `downto` 1 is a **null range** (0 `to` 0 is a legal range). Write a one-page summary on null ranges, including code examples. Is a null range treated as an ascending or descending range?

10.85 (**Loops, 45 min.) Investigate the following issues with loops, including code examples and the results of analysis and simulation:

a. Try to alter the loop parameter within a loop. What happens?

b. What is the type of the loop parameter?

c. Can the condition inside a loop depend on a loop parameter?

d. What happens in a `for` loop if the range is null?

e. Can you pass a loop parameter out of a procedure as a procedure parameter?

10.86 (Signals and variables, 30 min.) Write a summary on signals and variables, including code examples.

10.87 (Type conversion, 60 min.) There are some very subtle rules involving type conversion, [VHDL 93LRM7.3.5]. Does the following work? Explain the type conversion rules.

```
BV <= BIT_VECTOR("1111");
```

10.19 Bibliography

The definitive reference guide to VHDL is the IEEE VHDL LRM [IEEE, 1076-1993]. The LRM is initially difficult to read because it is concise and precise (the LRM is intended for tool builders and experienced tool users, not as a tutorial). The LRM does form a useful reference—as does a dictionary for serious users of any language. You might think of the LRM as a legal contract between you and the company that sells you software that is compliant with the standard. VHDL software uses the terminology of the LRM for error messages, so it is necessary to understand the terms and definitions of the LRM. The WAVES standard [IEEE 1029.1-1991] deals with the problems of interfacing VHDL testbenches to testers.

VHDL International maintains VIUF (VHDL International Users' Forum) Internet Services (`http:/www.vhdl.org`) and links to other groups working on VHDL including the IEEE synthesis packages, IEEE WAVES packages, and IEEE VITAL packages (see also Appendix A).

The frequently asked questions (FAQ) list for the VHDL newsgroup `comp.lang.vhdl` is a useful starting point (the list is archived at `gopher://kona.ee.pitt.edu/h0/NewsGroupArchives`). Information on character sets and the problems of exchanging information across national boundaries can be found at `ftp://watsun.cc.columbia.edu/kermit/charsets`.

10.20 References

Page numbers in brackets after the reference indicate the location in the chapter body.

IEEE 1029.1-1991. *IEEE Standard for Waveform and Vector Exchange (WAVES)*. IEEE Std 1029.1-1991. The Institute of Electrical and Electronics Engineers, Inc., New York. Available from The Institute of Electrical and Electronics Engineers, Inc., 345 East 47th Street, New York, NY 10017 USA.

IEEE 1076-1993. *IEEE Standard VHDL Language Reference Manual (ANSI)*. IEEE Std. 1076-1993. The Institute of Electrical and Electronics Engineers, Inc., New York. Available from The Institute of Electrical and Electronics Engineers, Inc., 345 East 47th Street, New York, NY 10017 USA. [p. 380]

IEEE 1076.2-1996. *Standard VHDL Language Mathematical Packages*. IEEE Ref. AD129-NYF. Approved by IEEE Standards Board on 19 September 1996. [p. 404].

ISO 8859-1. 1987 (E). Information Processing—8-bit single-byte coded graphic character sets—Part 1: Latin Alphabet No. 1. American National Standards Institute, Hackensack, NJ; 1987. Available from Sales Department, American National Standards Institute, 105-111 South State Street, Hackensack, NJ 07601 USA. [p. 391]

VERILOG HDL

In this chapter we look at the **Verilog** hardware description language. Gateway Design Automation developed Verilog as a simulation language. The use of the Verilog-XL simulator is discussed in more detail in Chapter 13. Cadence purchased Gateway in 1989 and, after some study, placed the Verilog language in the public domain. Open Verilog International (OVI) was created to develop the Verilog language as an IEEE standard. The definitive reference guide to the Verilog language is now the Verilog LRM, IEEE Std 1364-1995 [1995].[1] This does not mean that all Verilog simulators and tools adhere strictly to the IEEE Standard—we must abide by the reference manual for the software we are using. Verilog is a fairly simple language to learn, especially if you are familiar with the C programming language. In this chapter we shall concentrate on the features of Verilog applied to high-level design entry and synthesis for ASICs.

---

[1]Some of the material in this chapter is reprinted with permission from IEEE Std 1364-1995, © Copyright 1995 IEEE. All rights reserved.

11.1 A Counter

The following Verilog code models a "black box" that contains a 50 MHz clock (period 20 ns), counts from 0 to 7, resets, and then begins counting at 0 again:

```
`timescale 1ns/1ns   // Set the units of time to be nanoseconds.    //1
module counter;                                                     //2
  reg clock;         // Declare a reg data type for the clock.      //3
  integer count;     // Declare an integer data type for the count. //4
initial // Initialize things; this executes once at t=0.            //5
  begin                                                             //6
    clock = 0; count = 0;       // Initialize signals.              //7
    #340 $finish;               // Finish after 340 time ticks.     //8
  end                                                               //9
/* An always statement to generate the clock; only one statement
follows the always so we don't need a begin and an end. */          //10
always                                                              //11
    #10 clock = ~ clock; // Delay (10ns) is set to half the clock cycle.//12
/* An always statement to do the counting; this executes at the same
time (concurrently) as the preceding always statement. */           //13
always                                                              //14
  begin                                                             //15
    // Wait here until the clock goes from 1 to 0.                  //16
    @ (negedge clock);                                              //17
    // Now handle the counting.                                     //18
    if (count == 7)                                                 //19
      count = 0;                                                    //20
    else                                                            //21
      count = count + 1;                                            //22
    $display("time = ",$time," count = ", count);                   //23
  end                                                               //24
endmodule                                                           //25
```

Verilog **keywords** (reserved words that are part of the Verilog language) are shown in bold type in the code listings (but not in the text). References in this chapter such as [Verilog LRM 1.1] refer you to the IEEE Verilog LRM.

The following output is from the Cadence Verilog-XL simulator. This example includes the system input so you can see how the tool is run and when it is finished. Some of the banner information is omitted in the listing that follows to save space (we can use "quiet" mode using a '-q' flag, but then the version and other useful information is also suppressed):

```
> verilog counter.v
VERILOG-XL 2.2.1   Apr 17, 1996  11:48:18
     ... Banner information omitted here...
Compiling source file "counter.v"
Highest level modules:
```

```
counter


time =                   20 count =              1
time =                   40 count =              2
(... 12 lines omitted...)
time =                  300 count =              7
time =                  320 count =              0
L10 "counter.v": $finish at simulation time 340
223 simulation events
CPU time: 0.6 secs to compile + 0.2 secs to link + 0.0 secs in
simulation
End of VERILOG-XL 2.2.1   Apr 17, 1996  11:48:20
>
```

Here is the output of the VeriWell simulator from the console window (future
examples do not show all of the compiler output— just the model output):

```
Veriwell -k VeriWell.key -l VeriWell.log -s :counter.v
... banner information omitted ....
Memory Available: 0
Entering Phase I...
Compiling source file : :counter.v
The size of this model is [1%, 1%] of the capacity of the free version


Entering Phase II...
Entering Phase III...
No errors in compilation
Top-level modules:
    counter


C1> .
time =                   20 count =              1
time =                   40 count =              2
(... 12 lines omitted...)
time =                  300 count =              7
time =                  320 count =              0
Exiting VeriWell for Macintosh at time 340
0 Errors, 0 Warnings, Memory Used: 29468
Compile time = 0.6, Load time = 0.7, Simulation time = 4.7


Normal exit
Thank you for using VeriWell for Macintosh
```

11.2   Basics of the Verilog Language

A Verilog **identifier** [Verilog LRM2.7], including the names of variables, may contain any sequence of letters, digits, a dollar sign '$', and the underscore '_' symbol. The first character of an identifier must be a letter or underscore; it cannot be a dollar sign '$', for example. We cannot use characters such as '-' (hyphen), brackets, or '#' (for active-low signals) in Verilog names (escaped identifiers are an exception). The following is a shorthand way of saying the same thing:

```
identifier ::= simple_identifier | escaped_identifier
simple_identifier ::= [ a-zA-Z_ ][ a-zA-Z_$ ]
escaped_identifier ::=
    \ {Any_ASCII_character_except_white_space} white_space
white_space ::= space | tab | newline
```

(In the 1995 LRM the underscore '_' is missing from the first bracket.) If we think of '::=' as an equal sign, then the preceding "equation" defines the syntax of an identifier. Usually we use the Backus–Naur form (BNF) to write these equations. We also use the BNF to describe the syntax of VHDL. There is an explanation of the BNF in Appendix A. Verilog syntax definitions are given in Appendix B. In Verilog all names, including keywords and identifiers, are case-sensitive. Special commands for the simulator (a system task or a system function) begin with a dollar sign '$' [Verilog LRM 2.7]. Here are some examples of Verilog identifiers:

```
module identifiers;                                                    //1
/* Multiline comments in Verilog                                       //2
   look like C comments and // is OK in here. */                       //3
// Single-line comment in Verilog.                                     //4
reg legal_identifier,two__underscores;                                 //5
reg _OK,OK_,OK_$,OK_123,CASE_SENSITIVE, case_sensitive;                //6
reg \/clock ,\a*b ;        // Add white_space after escaped identifier. //7
//reg $_BAD,123_BAD;       // Bad names even if we declare them!        //8
initial begin                                                          //9
legal_identifier = 0;      // Embedded underscores are OK,             //10
two__underscores = 0;      // even two underscores in a row.           //11
_OK = 0;                   // Identifiers can start with underscore    //12
OK_ = 0;                   // and end with underscore.                 //13
OK$ = 0;                   // $ sign is OK, but beware foreign keyboards.//14
OK_123 =0;                 // Embedded digits are OK.                   //15
CASE_SENSITIVE = 0;        // Verilog is case-sensitive (unlike VHDL). //16
case_sensitive = 1;                                                    //17
\/clock = 0;               // An escaped identifier with \ breaks rules,//18
\a*b = 0;                  // but be careful to watch the spaces!      //19
$display("Variable CASE_SENSITIVE= %d",CASE_SENSITIVE);                //20
$display("Variable case_sensitive= %d",case_sensitive);                //21
$display("Variable \/clock = %d",\/clock );                            //22
$display("Variable \\a*b = %d",\a*b );                                 //23
```

```
end                                                          //24
endmodule                                                    //25
```

The following is the output from this model (future examples in this chapter list the simulator output directly after the Verilog code).

```
Variable CASE_SENSITIVE= 0
Variable case_sensitive= 1
Variable /clock = 0
Variable \a*b = 0
```

11.2.1   Verilog Logic Values

Verilog has a predefined logic-value system or **value set** [Verilog LRM 3.1] that uses four logic values: '0', '1', 'x', and 'z' (lowercase 'x' and lowercase 'z'). The value 'x' represents an uninitialized or an unknown logic value—an unknown value is either '1', '0', 'z', or a value that is in a state of change. The logic value 'z' represents a high-impedance value, which is usually treated as an 'x' value. Verilog uses a more complicated internal logic-value system in order to resolve conflicts between different drivers on the same node. This hidden logic-value system is useful for switch-level simulation, but for most ASIC simulation and synthesis purposes we do not need to worry about the internal logic-value system.

11.2.2   Verilog Data Types

There are several **data types** in Verilog—all except one need to be declared before we can use them. The two main data types are **nets** and **registers** [Verilog LRM 3.2]. Nets are further divided into several net types. The most common and important net types are: **wire** and **tri** (which are identical); **supply1** and **supply0** (which are equivalent to the positive and negative power supplies respectively). The wire data type (which we shall refer to as just wire from now on) is analogous to a wire in an ASIC. A wire cannot store or hold a value. A wire must be continuously driven by an assignment statement (see Section 11.5). The default initial value for a wire is 'z' [Verilog LRM3.6]. There are also **integer**, **time**, **event**, and **real** data types.

```
module declarations_1;                                       //1
wire pwr_good, pwr_on, pwr_stable; // Explicitly declare wires.  //2
integer i;        // 32-bit, signed (2's complement).        //3
time t;           // 64-bit, unsigned, behaves like a 64-bit reg.  //4
event e;          // Declare an event data type.             //5
real r;           // Real data type of implementation defined size.  //6
// An assign statement continuously drives a wire:           //7
assign pwr_stable = 1'b1; assign pwr_on = 1; // 1 or 1'b1    //8
assign pwr_good = pwr_on & pwr_stable;                       //9
initial begin                                                //10
i = 123.456;      // There must be a digit on either side    //11
r = 123456e-3;    // of the decimal point if it is present.  //12
```

```
t = 123456e-3;    // Time is rounded to 1 second by default.    //13
$display("i=%0g",i," t=%6.2f",t," r=%f",r);                     //14
#2 $display("TIME=%0d",$time," ON=",pwr_on,                     //15
   " STABLE=",pwr_stable," GOOD=",pwr_good);                    //16
$finish; end                                                    //17
endmodule                                                       //18

i=123 t=123.00 r=123.456000
TIME=2 ON=1 STABLE=1 GOOD=1
```

A **register** data type is declared using the keyword `reg` and is comparable to a variable in a programming language. On the LHS of an assignment a register data type (which we shall refer to as just `reg` from now on) is updated immediately and holds its value until changed again. The default initial value for a `reg` is `'x'`. We can transfer information directly from a `wire` to a `reg` as shown in the following code:

```
module declarations_2;                                          //1
reg Q, Clk; wire D;                                             //2
// Drive the wire (D):                                          //3
assign D = 1;                                                   //4
// At a +ve clock edge assign the value of wire D to the reg Q: //5
always @(posedge Clk) Q = D;                                    //6
initial Clk = 0; always #10 Clk = ~ Clk;                        //7
initial begin #50; $finish; end                                 //8
always begin                                                    //9
$display("T=%2g", $time," D=",D," Clk=",Clk," Q=",Q); #10; end  //10
endmodule                                                       //11

T= 0 D=z Clk=0 Q=x
T=10 D=1 Clk=1 Q=x
T=20 D=1 Clk=0 Q=1
T=30 D=1 Clk=1 Q=1
T=40 D=1 Clk=0 Q=1
```

We shall discuss assignment statements in Section 11.5. For now, it is important to recognize that a `reg` is not always equivalent to a hardware register, flip-flop, or latch. For example, the following code describes purely combinational logic:

```
module declarations_3;                                          //1
reg a,b,c,d,e;                                                  //2
initial begin                                                   //3
   #10; a = 0;b = 0;c = 0;d = 0; #10; a = 0;b = 1;c = 1;d = 0;  //4
   #10; a = 0;b = 0;c = 1;d = 1; #10; $stop;                    //5
end                                                             //6
always begin                                                    //7
   @(a or b or c or d) e = (a|b)&(c|d);                         //8
   $display("T=%0g",$time," e=",e);                             //9
end                                                             //10
endmodule                                                       //11

T=10 e=0
```

```
T=20 e=1
T=30 e=0
```

A single-bit `wire` or `reg` is a **scalar** (the default). We may also declare a `wire` or `reg` as a **vector** with a **range** of bits [Verilog LRM 3.3]. In some situations we may use implicit declaration for a scalar `wire`; it is the only data type we do not always need to declare. We must use explicit declaration for a vector `wire` or any `reg`. We may access (or **expand**) the range of bits in a vector one at a time, using a **bit-select**, or as a contiguous subgroup of bits (a continuous sequence of numbers— like a straight in poker) using a **part-select** [Verilog LRM 4.2]. The following code shows some examples:

```
module declarations_4;                                          //1
wire Data;                       // A scalar net of type wire.  //2
wire [31:0] ABus, DBus;          // Two 32-bit-wide vector wires: //3
// DBus[31] = leftmost = most-significant bit  = msb            //4
// DBus[0] = rightmost = least-significant bit = lsb            //5
// Notice the size declaration precedes the names.              //6
// wire [31:0] TheBus, [15:0] BigBus; // This is illegal.       //7
reg [3:0] vector;                // A 4-bit vector register.    //8
reg [4:7] nibble;                // msb index < lsb index is OK. //9
integer i;                                                      //10
initial begin                                                   //11
i = 1;                                                          //12
vector = 'b1010;                 // Vector without an index.    //13
nibble = vector;                 // This is OK too.             //14
#1; $display("T=%0g",$time," vector=", vector," nibble=", nibble); //15
#2; $display("T=%0g",$time," Bus=%b",DBus[15:0]);               //16
end                                                             //17
assign DBus [1] = 1;             // This is a bit-select.       //18
assign DBus [3:0] = 'b1111;      // This is a part-select.      //19
// assign DBus [0:3] = 'b1111;   // Illegal : wrong direction.  //20
endmodule                                                       //21
```

```
T=1 vector=10 nibble=10
T=3 Bus=zzzzzzzzzzzz1111
```

There are no multidimensional arrays in Verilog, but we may declare a **memory** data type as an **array** of registers [Verilog LRM 3.8]:

```
module declarations_5;                                          //1
reg [31:0] VideoRam [7:0]; // An 8-word by 32-bit wide memory.  //2
initial begin                                                   //3
VideoRam[1] = 'bxz; // We must specify an index for a memory.   //4
VideoRam[2] = 1;                                                //5
VideoRam[7] = VideoRam[VideoRam[2]]; // Need 2 clock cycles for this.//6
VideoRam[8] = 1; // Careful! the compiler won't complain about this! //7
// Verify what we entered:                                      //8
```

```
$display("VideoRam[0] is %b",VideoRam[0]);                    //9
$display("VideoRam[1] is %b",VideoRam[1]);                    //10
$display("VideoRam[2] is %b",VideoRam[2]);                    //11
$display("VideoRam[7] is %b",VideoRam[7]);                    //12
end                                                           //13
endmodule                                                     //14

VideoRam[0] is xxxxxxxxxxxxxxxxxxxxxxxxxxxxxxxx
VideoRam[1] is xxxxxxxxxxxxxxxxxxxxxxxxxxxxxxxz
VideoRam[2] is 00000000000000000000000000000001
VideoRam[7] is xxxxxxxxxxxxxxxxxxxxxxxxxxxxxxxz
```

We may also declare an **integer array** or **time array** in the same way as an array of reg, but there are no real arrays [Verilog LRM 3.9]:

```
module declarations_6;                                        //1
integer Number [1:100];    // Notice that size follows name    //2
time Time_Log [1:1000];    // - as in an array of reg.         //3
// real Illegal [1:10];    // Illegal. There are no real arrays.//4
endmodule                                                     //5
```

11.2.3 Other Wire Types

There are the following other Verilog wire types (rarely used in ASIC design) [Verilog LRM 3.7.2]:

- wand, wor, triand, and trior model wired logic. Wiring, or dotting, the outputs of two gates generates a logic function (in emitter-coupled logic, ECL, or in an EPROM, for example). This is one area in which the logic values 'z' and 'x' are treated differently.
- tri0 and tri1 model resistive connections to VSS or VDD.
- trireg is like a wire but associates some capacitance with the net, so it can model charge storage.

There are also other keywords that may appear in declarations:

- scalared and vectored are properties of vectors [Verilog LRM 3.3.2].
- small, medium, and large model the charge strength of trireg connections [Verilog LRM 7].

11.2.4 Numbers

Constant numbers are integer or real constants [Verilog LRM 2.5]. **Integer constants** are written as

$$\text{width'radix value}$$

where width and radix are optional. The **radix** (or base) indicates the type of number: **decimal** (d or D), **hex** (h or H), **octal** (o or O), or **binary** (b or B). A number may be **sized** or **unsized**. The length of an unsized number is implementation dependent.

We can use '1' and '0' as numbers since they cannot be identifiers, but we must write 1'bx and 1'bz for 'x' and 'z'. A number may be declared as a **parameter** [Verilog LRM 3.10]. A parameter assignment belongs inside a module declaration and has **local scope** [Verilog LRM3.11]. **Real constants** are written using decimal (100.0) or scientific notation (1e2) and follow IEEE Std 754-1985 for double-precision floating-point numbers. Reals are rounded to the nearest integer, ties (numbers that end in .5) round away from zero [Verilog LRM 3.9.2], but not all implementations follow this rule (the output from the following code is from VeriWell, which rounds ties toward zero for negative integers).

```
module constants;                                                   //1
parameter H12_UNSIZED = 'h 12;         // Unsized hex 12 = decimal 18. //2
parameter H12_SIZED = 6'h 12;          // Sized hex 12 = decimal 18.   //3
// Note: a space between base and value is OK.                      //4
// Note: '' (single apostrophes) are not the same as the ' character.//5
parameter D42 = 8'B0010_1010;          // bin 101010 = dec 42          //6
// OK to use underscores to increase readability.                   //7
parameter D123 = 123;                  // Unsized decimal (the default).//8
parameter D63 = 8'o 77;                // Sized octal, decimal 63.     //9
// parameter ILLEGAL = 1'o9;           // No 9's in octal numbers!    //10
// A = 'hx and B = 'ox assume a 32 bit width.                       //11
parameter A = 'h x, B = 'o x, C = 8'b x, D = 'h z, E = 16'h ????;   //12
// Note the use of ? instead of z, 16'h ???? is the same as 16'h zzzz.//13
// Also note the automatic extension to a width of 16 bits.         //14
reg [3:0] B0011,Bxxx1,Bzzz1; real R1,R2,R3; integer I1,I3,I_3;      //15
parameter BXZ = 8'b1x0x1z0z;                                        //16
initial begin                                                       //17
B0011 = 4'b11; Bxxx1 = 4'bx1; Bzzz1 = 4'bz1;    // Left padded.      //18
R1 = 0.1e1; R2 = 2.0; R3 = 30E-01;        // Real numbers.          //19
I1 = 1.1; I3 = 2.5; I_3 = -2.5;           // IEEE rounds away from 0. //20
end                                                                 //21
initial begin #1;                                                   //22
$display                                                            //23
("H12_UNSIZED, H12_SIZED (hex) = %h, %h",H12_UNSIZED, H12_SIZED);   //24
$display("D42 (bin) = %b",D42," (dec) = %d",D42);                   //25
$display("D123 (hex) = %h",D123," (dec) = %d",D123);               //26
$display("D63 (oct) = %o",D63);                                     //27
$display("A (hex) = %h",A," B (hex) = %h",B);                       //28
$display("C (hex) = %h",C," D (hex) = %h",D," E (hex) = %h",E);     //29
$display("BXZ (bin) = %b",BXZ," (hex) = %h",BXZ);                   //30
$display("B0011, Bxxx1, Bzzz1 (bin) = %b, %b, %b",B0011,Bxxx1,Bzzz1);//31
$display("R1, R2, R3 (e, f, g) = %e, %f, %g", R1, R2, R3);          //32
$display("I1, I3, I_3 (d) = %d, %d, %d", I1, I3, I_3);              //33
end                                                                 //34
endmodule                                                           //35

H12_UNSIZED, H12_SIZED (hex) = 00000012, 12
D42 (bin) = 00101010 (dec) =  42
D123 (hex) = 0000007b (dec) =        123
```

```
D63 (oct) = 077
A (hex) = xxxxxxxx B (hex) = xxxxxxxx
C (hex) = xx D (hex) = zzzzzzzz E (hex) = zzzz
BXZ (bin) = 1x0x1z0z (hex) = XZ
B0011, Bxxx1, Bzzz1 (bin) = 0011, xxx1, zzz1
R1, R2, R3 (e, f, g) = 1.000000e+00, 2.000000, 3
I1, I3, I_3 (d) =              1,         3,        -2
```

11.2.5   Negative Numbers

Integer numbers are **signed** (two's complement) or **unsigned**. The following example illustrates the handling of negative constants [Verilog LRM 3.2.2, 4.1.3]:

```
module negative_numbers;                                                 //1
parameter PA = -12, PB = -'d12, PC = -32'd12, PD = -4'd12;               //2
integer IA , IB , IC , ID ; reg [31:0] RA , RB , RC , RD ;               //3
initial begin #1;                                                        //4
IA = -12; IB = -'d12; IC = -32'd12; ID = -4'd12;                         //5
RA = -12; RB = -'d12; RC = -32'd12; RD = -4'd12; #1;                     //6
$display("           parameter     integer    reg[31:0]");               //7
$display ("-12       =",PA,IA,,,RA);                                     //8
$displayh("           ",,,,PA,,,,IA,,,,,RA);                             //9
$display ("-'d12     =",,PB,IB,,,RB);                                    //10
$displayh("           ",,,,PB,,,,IB,,,,,RB);                             //11
$display ("-32'd12 =",,PC,IC,,,RC);                                      //12
$displayh("           ",,,,PC,,,,IC,,,,,RC);                             //13
$display ("-4'd12   =",,,,,,,,,,PD,ID,,,RD);                             //14
$displayh("           ",,,,,,,,,,,PD,,,,ID,,,,,,RD);                     //15
end                                                                      //16
endmodule                                                                //17
```

```
          parameter    integer    reg[31:0]
-12     =        -12        -12   4294967284
          fffffff4   fffffff4     fffffff4
-'d12   = 4294967284        -12   4294967284
          fffffff4   fffffff4     fffffff4
-32'd12 = 4294967284        -12   4294967284
          fffffff4   fffffff4     fffffff4
-4'd12  =          4        -12   4294967284
                   4   fffffff4     fffffff4
```

Verilog only "keeps track" of the sign of a negative constant if it is (1) assigned to an `integer` or (2) assigned to a `parameter` without using a base (essentially the same thing). In other cases (even though the bit representations may be identical to the signed number—hexadecimal `ffffff4` in the previous example), a negative constant is treated as an unsigned number. Once Verilog "loses" the sign, keeping track of signed numbers becomes your responsibility (see also Section 11.3.1).

11.2.6   Strings

The code listings in this book use Courier font. The ISO/ANSI standard for the ASCII code defines the characters, but not the appearance of the graphic symbol in any particular font. The confusing characters are the quote and accent characters:

```
module characters; /*                                              //1
" is ASCII 34 (hex 22), double quote.                              //2
' is ASCII 39 (hex 27), tick or apostrophe.                        //3
/ is ASCII 47 (hex 2F), forward slash.                             //4
\ is ASCII 92 (hex 5C), back slash.                                //5
` is ASCII 96 (hex 60), accent grave.                              //6
| is ASCII 124 (hex 7C), vertical bar.                             //7
There are no standards for the graphic symbols for codes above 128. //8
´ is 171 (hex AB), accent acute in almost all fonts.               //9
" is 210 (hex D2), open  double quote, like 66 (in some fonts).    //10
" is 211 (hex D3), close double quote, like 99 (in some fonts).    //11
' is 212 (hex D4), open  single quote, like  6 (in some fonts).    //12
' is 213 (hex D5), close single quote, like  9 (in some fonts).    //13
*/ endmodule                                                       //14
```

Here is an example showing the use of **string constants** [Verilog LRM 2.6]:

```
module text;                                                       //1
parameter A_String = "abc"; // string constant, must be on one line //2
parameter Say = "Say \"Hey!\"";                                    //3
// use escape quote \" for an embedded quote                       //4
parameter Tab = "\t";              // tab character                //5
parameter NewLine = "\n";          // newline character            //6
parameter BackSlash = "\\";        // back slash                   //7
parameter Tick = "\047";           // ASCII code for tick in octal //8
// parameter Illegal = "\500";     // illegal - no such ASCII code //9
initial begin                                                      //10
$display("A_String(str) = %s ",A_String," (hex) = %h ",A_String);  //11
$display("Say = %s ",Say," Say \"Hey!\"");                         //12
$display("NewLine(str) = %s ",NewLine," (hex) = %h ",NewLine);     //13
$display("\\(str) = %s ",BackSlash," (hex) = %h ",BackSlash);      //14
$display("Tab(str) = %s ",Tab," (hex) = %h ",Tab,"1 newline...");  //15
$display("\n");                                                    //16
$display("Tick(str) = %s ",Tick," (hex) = %h ",Tick);             //17
#1.23; $display("Time is %t", $time);                              //18
end                                                                //19
endmodule                                                          //20
```

```
A_String(str) = abc  (hex) = 616263
Say = Say \"Hey!\"  Say "Hey!"
NewLine(str) = \n  (hex) = 0a
\(str) = \\  (hex) = 5c
```

```
Tab(str) = \t   (hex) = 09 1 newline...


Tick(str) = '   (hex) = 27
Time is                      1
```

Instead of parameters you may use a **define directive** that is a **compiler directive**, and not a statement [Verilog LRM 16]. The define directive has **global scope**:

```
module define;                                                      //1
`define G_BUSWIDTH 32 // Bus width parameter (G_ for global).       //2
/* Note: there is no semicolon at end of a compiler directive. The
character ` is ASCII 96 (hex 60), accent grave, it slopes down from
left to right. It is not the tick or apostrophe character ' (ASCII 39
or hex 27)*/                                                        //3
wire [`G_BUSWIDTH:0]MyBus; // A 32-bit bus.                         //4
endmodule                                                           //5
```

11.3 Operators

An expression uses any of the three types of operators: unary operators, binary operators, and a single ternary operator [Verilog LRM 4.1]. The Verilog operators are similar to those in the C programming language—except there is no autoincrement (++) or autodecrement (--) in Verilog. Table 11.1 shows the operators in their (increasing) order of precedence and Table 11.2 shows the unary operators. Here is an example that illustrates the use of the Verilog operators:

```
module operators;                                                  //1
parameter A10xz = {1'b1,1'b0,1'bx,1'bz}; // Concatenation and      //2
parameter A01010101 = {4{2'b01}};    // replication, illegal for real.//3
// Arithmetic operators: +, -, *, /, and modulus %                 //4
parameter A1 = (3+2) %2; // The sign of a % b is the same as sign of a.//5
// Logical shift operators: << (left), >> (right)                  //6
parameter A2 = 4 >> 1; parameter A4 = 1 << 2;    // Note: zero fill.//7
// Relational operators: <, <=, >, >=                              //8
initial if (1 > 2) $stop;                                          //9
// Logical operators: ! (negation), && (and), || (or)             //10
parameter B0 = !12; parameter B1 = 1 && 2;                        //11
reg [2:0] A00x; initial begin A00x = 'b111; A00x = !2'bx1; end     //12
parameter C1 = 1 || (1/0); /* This may or may not cause an        //13
error: the short-circuit behavior of && and || is undefined. An   //14
evaluation including && or || may stop when an expression is known //15
to be true or false. */                                           //16
// == (logical equality), != (logical inequality)                 //17
```

TABLE 11.1   Verilog operators (in increasing order of precedence).

```
?: (conditional) [legal for real; associates right to left (others associate left to right)]
|| (logical or) [A smaller operand is zero-filled from its msb (0-fill); legal for real]
&& (logical and)[0-fill, legal for real]
| (bitwise or) ~| (bitwise nor) [0-fill]
^ (bitwise xor) ^~ ~^ (bitwise xnor, equivalence) [0-fill]
& (bitwise and) ~& (bitwise nand) [0-fill]
== (logical) != (logical) === (case) !== (case) [0-fill, logical versions are legal for real]
< (lt) <= (lt or equal) > (gt) >= (gt or equal) [0-fill, all arelegal for real]
<< (shift left) >> (shift right) [zero fill; no -ve shifts; shift by x or z results in unknown]
+ (addition) - (subtraction) [if any bit is x or z for + - * / % then entire result is unknown]
* (multiply) / (divide) % (modulus) [integer divide truncates fraction; + - * / legal for real]
Unary operators: ! ~ & ~& | ~| ^ ~^ ^~ + - [see Table 11.2 for precedence]
```

TABLE 11.2   Verilog unary operators.

Operator	Name	Examples
!	logical negation	!123 is 'b0 [0, 1, or x for ambiguous; legal for real]
~	bitwise unary negation	~1'b10xz is 1'b01xx
&	unary reduction and	& 4'b1111 is 1'b1, & 2'bx1 is 1'bx, & 2'bz1 is 1'bx
~&	unary reduction nand	~& 4'b1111 is 1'b0, ~& 2'bx1 is 1'bx
		unary reduction or	Note:
~|	unary reduction nor	Reduction is performed left (first bit) to right
^	unary reduction xor	Beware of the non-associative reduction operators
~^ ^~	unary reduction xnor	z is treated as x for all unary operators
+	unary plus	+2'bxz is +2'bxz [+m is the same as m; legal for real]
-	unary minus	-2'bxz is x [-m is unary minus m; legal for real]

```
parameter Ax = (1==1'bx); parameter Bx = (1'bx!=1'bz);        //18
parameter D0 = (1==0); parameter D1 = (1==1);                 //19
// === case equality, !== (case inequality)                   //20
// The case operators only return true (1) or false (0).      //21
parameter E0 = (1===1'bx); parameter E1 = 4'b01xz === 4'b01xz; //22
parameter F1 = (4'bxxxx === 4'bxxxx);                         //23
```

```
// Bitwise logical operators:                                          //24
// ~ (negation), & (and), | (inclusive or),                           //25
// ^ (exclusive or), ~^ or ^~ (equivalence)                           //26
parameter A00 = 2'b01 & 2'b10;                                        //27
// Unary logical reduction operators:                                  //28
// & (and), ~& (nand), | (or), ~| (nor),                             //29
// ^ (xor), ~^ or ^~ (xnor)                                           //30
parameter G1= & 4'b1111;                                              //31
// Conditional expression f = a ? b : c [if (a) then f=b else f=c]    //32
// if a=(x or z), then (bitwise) f=0 if b=c=0, f=1 if b=c=1, else f=x //33
reg H0, a, b, c; initial begin a=1; b=0; c=1; H0=a?b:c; end           //34
reg[2:0] J01x, Jxxx, J01z, J011;                                      //35
initial begin Jxxx = 3'bxxx; J01z = 3'b01z; J011 = 3'b011;           //36
J01x = Jxxx ? J01z : J011; end       // A bitwise result.             //37
initial begin #1;                                                     //38
$display("A10xz=%b",A10xz,"  A01010101=%b",A01010101);               //39
$display("A1=%0d",A1,"  A2=%0d",A2,"  A4=%0d",A4);                   //40
$display("B1=%b",B1,"  B0=%b",B0,"  A00x=%b",A00x);                  //41
$display("C1=%b",C1,"  Ax=%b",Ax,"  Bx=%b",Bx);                     //42
$display("D0=%b",D0,"  D1=%b",D1);                                   //43
$display("E0=%b",E0,"  E1=%b",E1,"  F1=%b",F1);                      //44
$display("A00=%b",A00,"  G1=%b",G1,"  H0=%b",H0);                    //45
$display("J01x=%b",J01x); end                                         //46
endmodule                                                             //47
```

```
A10xz=10xz   A01010101=01010101
A1=1    A2=2   A4=4
B1=1    B0=0   A00x=00x
C1=1    Ax=x   Bx=x
D0=0    D1=1
E0=0    E1=1   F1=1
A00=00   G1=1   H0=0
J01x=01x
```

11.3.1 Arithmetic

Arithmetic operations on n-bit objects are performed modulo 2^n in Verilog,

```
module modulo; reg [2:0] Seven;                                       //1
initial begin                                                         //2
#1 Seven = 7; #1 $display("Before=", Seven);                         //3
#1 Seven = Seven + 1; #1 $display("After =", Seven);                 //4
end                                                                   //5
endmodule                                                             //6
```

```
Before=7
After =0
```

Arithmetic operations in Verilog (addition, subtraction, comparison, and so on) on vectors (`reg` or `wire`) are predefined (Tables 11.1 and 11.2 show which operators are legal for `real`). This is a very important difference for ASIC designers from the situation in VHDL. However, there are some subtleties with Verilog arithmetic and negative numbers that are illustrated by the following example (based on an example in the LRM [Verilog LRM4.1.3]):

```
module LRM_arithmetic;                                                  //1
integer IA, IB, IC, ID, IE; reg [15:0] RA, RB, RC;                      //2
initial begin                                                           //3
IA = -4'd12;           RA =  IA / 3; // reg is treated as unsigned. //4
RB = -4'd12;           IB =  RB / 3; //                                  //5
IC = -4'd12 / 3;       RC = -12 / 3; // real is treated as signed       //6
ID =    -12 / 3;       IE =  IA / 3; // (two's complement).              //7
end                                                                     //8
initial begin #1;                                                       //9
$display("              hex     default");                              //10
$display("IA = -4'd12     = %h%d",IA,IA);                               //11
$display("RA = IA / 3     =    %h     %d",RA,RA);                        //12
$display("RB = -4'd12     =    %h     %d",RB,RB);                        //13
$display("IB = RB / 3     = %h%d",IB,IB);                                //14
$display("IC = -4'd12 / 3 = %h%d",IC,IC);                                //15
$display("RC = -12 / 3    =    %h     %d",RC,RC);                        //16
$display("ID = -12 / 3    = %h%d",ID,ID);                                //17
$display("IE =  IA / 3    = %h%d",IE,IE);                                //18
end                                                                     //19
endmodule                                                               //20
```

```
                   hex     default
IA = -4'd12     = fffffff4       -12
RA = IA / 3     =     fffc     65532
RB = -4'd12     =     fff4     65524
IB = RB / 3     = 00005551     21841
IC = -4'd12 / 3 = 55555551 1431655761
RC = -12 / 3    =     fffc     65532
ID = -12 / 3    = fffffffc        -4
IE =  IA / 3    = fffffffc        -4
```

We might expect the results of all these divisions to be $-4 = -12/3$. For integer assignments, the results are correctly signed (`ID` and `IE`). Hex `fffc` (decimal 65532) is the 16-bit two's complement of -4, so `RA` and `RC` are also correct if we keep track of the signs ourselves. The integer result `IB` is incorrect because Verilog treats `RB` as an unsigned number. Verilog also treats `-4'd12` as an unsigned number in the calculation of `IC`. Once Verilog "loses" a sign, it cannot get it back (see also Section 11.2.5).

11.4 Hierarchy

The **module** is the basic unit of code in the Verilog language [Verilog LRM 12.1],

```
module holiday_1(sat, sun, weekend);                              //1
    input sat, sun; output weekend;                               //2
    assign weekend = sat | sun;                                   //3
endmodule                                                         //4
```

We do not have to explicitly declare the scalar wires: saturday, sunday, weekend because, since these wires appear in the module interface, they must be declared in an input, output, or inout statement and are thus implicitly declared. The **module interface** provides the means to interconnect two Verilog modules using **ports** [Verilog LRM 12.3]. Each port must be explicitly declared as one of **input**, **output**, or **inout**. Table 11.3 shows the characteristics of ports. Notice that a reg cannot be an input port or an inout port. This is to stop us trying to connect a reg to another reg that may hold a different value.

TABLE 11.3 Verilog ports.

Verilog port Characteristics	input	output	inout
	wire (or other net)	**reg** or **wire** (or other net) We *can* read an output port inside a module	**wire** (or other net)

Within a module we may **instantiate** other modules, but we cannot declare other modules. Ports are linked using **named association** or **positional association**,

```
`timescale 100s/1s // Units are 100 seconds with precision of 1s.   //1
module life; wire [3:0] n; integer days;                            //2
    wire wake_7am, wake_8am; // Wake at 7 on weekdays else at 8.     //3
    assign n = 1 + (days % 7); // n is day of the week (1-7)         //4
always@(wake_8am or wake_7am)                                       //5
    $display("Day=",n," hours=%0d ",($time/36)%24,"  8am = ",       //6
      wake_8am," 7am = ",wake_7am,"  m2.weekday = ", m2.weekday);    //7
    initial days = 0;                                               //8
    initial begin #(24*36*10);$finish; end // Run for 10 days.      //9
    always #(24*36) days = days + 1; // Bump day every 24hrs.       //10
    rest m1(n, wake_8am); // Module instantiation.                  //11
// Creates a copy of module rest with instance name m1,             //12
// ports are linked using positional notation.                     //13
    work m2(.weekday(wake_7am), .day(n));                           //14
// Creates a copy of module work with instance name m2,            //15
```

```
// Ports are linked using named association.              //16
endmodule                                                 //17

module rest(day, weekend); // Module definition.          //1
// Notice the port names are different from the parent.    //2
   input [3:0] day; output weekend; reg weekend;          //3
   always begin #36 weekend = day > 5; end // Need a delay here.  //4
endmodule                                                 //5

module work(day, weekday);                                //1
   input [3:0] day; output weekday; reg weekday;          //2
   always begin #36 weekday = day < 6; end // Need a delay here.  //3
endmodule                                                 //4

Day= 1 hours=0   8am = 0   7am = 0   m2.weekday = 0
Day= 1 hours=1   8am = 0   7am = 1   m2.weekday = 1
Day= 6 hours=1   8am = 1   7am = 0   m2.weekday = 0
Day= 1 hours=1   8am = 0   7am = 1   m2.weekday = 1
```

The port names in a module definition and the port names in the parent module may be different. We can **associate** (link or map) ports using the same order in the instantiating statement as we use in the module definition—such as instance m1 in module life. Alternatively we can associate the ports by naming them—such as instance m2 in module life (using a period '.' before the port name that we declared in the module definition). Identifiers in a module have local scope. If we want to refer to an identifier outside a module, we use a **hierarchical name** [Verilog LRM12.4] such as m1.weekend or m2.weekday (as in module life), for example. The compiler will first search downward (or inward) then upward (outward) to resolve a hierarchical name [Verilog LRM 12.4–12.5].

11.5 Procedures and Assignments

A Verilog **procedure** [Verilog LRM 9.9] is an always or initial statement, a task, or a function. The statements within a sequential block (statements that appear between a begin and an end) that is part of a procedure execute sequentially in the order in which they appear, but the procedure executes concurrently with other procedures. This is a fundamental difference from computer programming languages. Think of each procedure as a microprocessor running on its own and at the same time as all the other microprocessors (procedures). Before I discuss procedures in more detail, I shall discuss the two different types of assignment statements:

- *continuous assignments* that appear outside procedures
- *procedural assignments* that appear inside procedures

To illustrate the difference between these two types of assignments, consider again the example used in Section 11.4:

```
module holiday_1(sat, sun, weekend);                                    //1
   input sat, sun; output weekend;                                      //2
   assign weekend = sat | sun; // Assignment outside a procedure.       //3
endmodule                                                               //4
```

We can change weekend to a reg instead of a wire, but then we must declare weekend and use a procedural assignment (inside a procedure—an always statement, for example) instead of a continuous assignment. We also need to add some delay (one time tick in the example that follows); otherwise the computer will never be able to get out of the always procedure to execute any other procedures:

```
module holiday_2(sat, sun, weekend);                                    //1
   input sat, sun; output weekend; reg weekend;                         //2
   always #1 weekend = sat | sun; // Assignment inside a procedure.     //3
endmodule                                                               //4
```

We shall cover the continuous assignment statement in the next section, which is followed by an explanation of sequential blocks and procedural assignment statements. Here is some skeleton code that illustrates where we may use these assignment statements:

```
module assignments                                                      //1
//... Continuous assignments go here.                                   //2
always // beginning of a procedure                                      //3
   begin // beginning of sequential block                               //4
   //... Procedural assignments go here.                                //5
   end                                                                  //6
endmodule                                                               //7
```

Table 11.4 at the end of Section 11.6 summarizes assignment statements, including two more forms of assignment—you may want to look at this table now.

11.5.1   Continuous Assignment Statement

A **continuous assignment statement** [Verilog LRM 6.1] assigns a value to a wire in a similar way that a real logic gate drives a real wire,

```
module assignment_1();                                                  //1
wire pwr_good, pwr_on, pwr_stable; reg Ok, Fire;                        //2
assign pwr_stable = Ok & (!Fire);                                       //3
assign pwr_on = 1;                                                      //4
assign pwr_good = pwr_on & pwr_stable;                                  //5
initial begin Ok = 0; Fire = 0; #1 Ok = 1; #5 Fire = 1; end             //6
initial begin $monitor("TIME=%0d",$time," ON=",pwr_on, " STABLE=",      //7
     pwr_stable," OK=",Ok," FIRE=",Fire," GOOD=",pwr_good);             //8
```

```
   #10 $finish; end                                              //9
endmodule                                                        //10
```

```
TIME=0 ON=1 STABLE=0 OK=0 FIRE=0 GOOD=0
TIME=1 ON=1 STABLE=1 OK=1 FIRE=0 GOOD=1
TIME=6 ON=1 STABLE=0 OK=1 FIRE=1 GOOD=0
```

The assignment statement in this next example models a three-state bus:

```
module assignment_2; reg Enable; wire [31:0] Data;              //1
/* The following single statement is equivalent to a declaration and
continuous assignment. */                                       //2
wire [31:0] DataBus = Enable ? Data : 32'bz;                    //3
assign Data = 32'b10101101101011101111000010100001;            //4
   initial begin                                                //5
      $monitor("Enable=%b DataBus=%b ", Enable, DataBus);       //6
      Enable = 0; #1; Enable = 1; #1; end                       //7
endmodule                                                        //8
```

```
Enable = 0 DataBus =zzzzzzzzzzzzzzzzzzzzzzzzzzzzzzzz
Enable = 1 DataBus =10101101101011101111000010100001
```

11.5.2   Sequential Block

A **sequential block** [Verilog LRM 9.8] is a group of statements between a **begin** and an **end**. We may declare new variables within a sequential block, but then we must name the block. A sequential block is considered a statement, so that we may nest sequential blocks.

A sequential block may appear in an **always statement** [Verilog LRM9.9.2], in which case the block executes repeatedly. In contrast, an **initial statement** [Verilog LRM9.9.1] executes only once, so a sequential block within an initial statement only executes once—at the beginning of a simulation. It does not matter where the initial statement appears—it still executes first. Here is an example:

```
module always_1; reg Y, Clk;                                    //1
always // Statements in an always statement execute repeatedly: //2
begin: my_block                 // Start of sequential block.    //3
   @(posedge Clk) #5 Y = 1;     // At +ve edge set Y=1,          //4
   @(posedge Clk) #5 Y = 0;     // at the NEXT +ve edge set Y=0. //5
end                             // End of sequential block.      //6
always #10 Clk = ~ Clk;         // We need a clock.              //7
initial Y = 0;                  // These initial statements execute //8
initial Clk = 0;                // only once, but first.         //9
initial $monitor("T=%2g",$time," Clk=",Clk," Y=",Y);           //10
initial #70 $finish;                                             //11
endmodule                                                        //12
```

```
T= 0  Clk=0  Y=0
T=10  Clk=1  Y=0
T=15  Clk=1  Y=1
```

```
T=20   Clk=0   Y=1
T=30   Clk=1   Y=1
T=35   Clk=1   Y=0
T=40   Clk=0   Y=0
T=50   Clk=1   Y=0
T=55   Clk=1   Y=1
T=60   Clk=0   Y=1
```

11.5.3 Procedural Assignments

A **procedural assignment** [Verilog LRM 9.2] is similar to an assignment statement
in a computer programming language such as C. In Verilog the value of an expression on the RHS of an assignment within a procedure (a procedural assignment)
updates a `reg` (or memory element) on the LHS. In the absence of any *timing
controls* (see Section 11.6), the `reg` is updated immediately when the statement executes. The `reg` holds its value until changed by another procedural assignment. Here
is the BNF definition:

```
blocking_assignment ::= reg_lvalue = [delay_or_event_control] expression
```

(Notice this BNF definition is for a *blocking* assignment—a type of procedural
assignment—see Section 11.6.4.) Here is an example of a procedural assignment
(notice that a `wire` can only appear on the RHS of a procedural assignment):

```
module procedural_assign; reg Y, A;                                        //1
always @(A)                                                                //2
  Y = A; // Procedural assignment.                                         //3
initial begin A=0; #5; A=1; #5; A=0; #5; $finish; end                      //4
initial $monitor("T=%2g",$time,,"A=",A,,,"Y=",Y);                          //5
endmodule                                                                  //6
```

```
T= 0 A=0   Y=0
T= 5 A=1   Y=1
T=10 A=0   Y=0
```

11.6 Timing Controls and Delay

The statements within a sequential block are executed in order, but, in the absence
of any delay, they all execute at the same simulation time—the current **time step**. In
reality there are delays that are modeled using a timing control.

11.6.1 Timing Control

A **timing control** is either a delay control or an event control [Verilog LRM 9.7]. A
delay control delays an assignment by a specified amount of time. A **timescale**

compiler directive is used to specify the units of time followed by the precision used to calculate time expressions,

```
`timescale 1ns/10ps // Units of time are ns. Round times to 10 ps.
```

Time units may only be `s`, `ns`, `ps`, or `fs` and the multiplier must be 1, 10, or 100. We can delay an assignment in two different ways:

- Sample the RHS immediately and then delay the assignment to the LHS.
- Wait for a specified time and then assign the value of the RHS to the LHS.

Here is an example of the first alternative (an **intra-assignment delay**):

```
x = #1 y;          // intra-assignment delay
```

The second alternative is **delayed assignment**:

```
#1 x = y;          // delayed assignment
```

These two alternatives are not the same. The intra-assignment delay is equivalent to the following code:

```
begin              // Equivalent to intra-assignment delay.
  hold = y;        // Sample and hold y immediately.
  #1;              // Delay.
  x = hold;        // Assignment to x. Overall same as x = #1 y.
end
```

In contrast, the delayed assignment is equivalent to a delay followed by an assignment as follows:

```
begin              // Equivalent to delayed assignment.
  #1;              // Delay.
  x = y;           // Assign y to x. Overall same as #1 x = y.
end
```

The other type of timing control, an **event control**, delays an assignment until a specified event occurs. Here is the formal definition:

```
event_control ::= @ event_identifier | @ (event_expression)

event_expression ::= expression | event_identifier
  | posedge expression | negedge expression
  | event_expression or event_expression
```

(Notice there are two different uses of `'or'` in this simplified BNF definition—the last one, in bold, is part of the Verilog language, a keyword.) A positive edge (denoted by the keyword `posedge`) is a transition from `'0'` to `'1'` or `'x'`, or a transition from `'x'` to `'1'`. A negative edge (`negedge`) is a transition from `'1'` to `'0'` or

'x', or a transition from 'x' to '0'. Transitions to or from 'z' do not count. Here
are examples of event controls:

```
module delay_controls; reg X, Y, Clk, Dummy;                         //1
always #1 Dummy=!Dummy; // Dummy clock, just for graphics.           //2
// Examples of delay controls:                                      //3
always begin #25 X=1;#10 X=0;#5; end                                //4
// An event control:                                                //5
always @(posedge Clk) Y=X; // Wait for +ve clock edge.              //6
always #10 Clk = !Clk; // The real clock.                           //7
initial begin Clk = 0;                                              //8
   $display("T   Clk X Y");                                         //9
   $monitor("%2g",$time,,,Clk,,,,X,,Y);                             //10
   $dumpvars;#100 $finish; end                                      //11
endmodule                                                           //12
```

```
T    Clk X Y
 0    0    x x
10    1    x x
20    0    x x
25    0    1 x
30    1    1 1
35    1    0 1
40    0    0 1
50    1    0 0
60    0    0 0
65    0    1 0
70    1    1 1
75    1    0 1
80    0    0 1
90    1    0 0
```

The dummy clock in delay_controls helps in the graphical waveform display
of the results (it provides a one-time-tick timing grid when we zoom in, for example). Figure 11.1 shows the graphical output from the Waves viewer in VeriWell
(white is used to represent the initial unknown values). The assignment statements to
'X' in the always statement repeat (every $25 + 10 + 5 = 40$ time ticks).

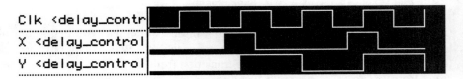

FIGURE 11.1 Output
from the module
delay_controls.

Events can be declared (as **named events**), triggered, and detected as follows:

```
module show_event;                                              //1
reg clock;                                                      //2
event event_1, event_2; // Declare two named events.            //3
always @(posedge clock) -> event_1; // Trigger event_1.         //4
always @ event_1                                                //5
begin $display("Strike 1!!"); -> event_2; end // Trigger event_2. //6
always @ event_2 begin $display("Strike 2!!");                  //7
$finish; end // Stop on detection of event_2.                   //8
always #10 clock = ~ clock; // We need a clock.                 //9
initial clock = 0;                                              //10
endmodule                                                       //11
```

```
Strike 1!!
Strike 2!!
```

11.6.2   Data Slip

Consider this model for a shift register and the simulation output that follows:

```
module data_slip_1 (); reg Clk, D, Q1, Q2;                      //1
/*********** bad sequential logic below **************/         //2
always @(posedge Clk)  Q1 = D;                                  //3
always @(posedge Clk)  Q2 = Q1; // Data slips here!             //4
/*********** bad sequential logic above **************/         //5
initial begin Clk = 0; D = 1; end always #50 Clk = ~Clk;        //6
initial begin $display("t    Clk D Q1 Q2");                     //7
$monitor("%3g",$time,,Clk,,,,D,,Q1,,,Q2); end                   //8
initial #400 $finish; // Run for 8 cycles.                      //9
initial $dumpvars;                                              //10
endmodule                                                       //11
```

```
t    Clk D Q1 Q2
  0  0   1 x  x
 50  1   1 1  1
100  0   1 1  1
150  1   1 1  1
200  0   1 1  1
250  1   1 1  1
300  0   1 1  1
350  1   1 1  1
```

The first clock edge at t = 50 causes Q1 to be updated to the value of D at the clock edge (a '1'), and at the same time Q2 is updated to this new value of Q1. The data, D, has passed through both always statements. We call this problem **data slip**.

If we include delays in the `always` statements (labeled 3 and 4) in the preceding example, like this—

```
always @(posedge Clk)  Q1 = #1 D;  // The delays in the assignments  //3
always @(posedge Clk)  Q2 = #1 Q1; // fix the data slip.              //4
```

—we obtain the correct output:

```
t    Clk D Q1 Q2
  0 0    1 x  x
 50 1    1 x  x
 51 1    1 1  x
100 0    1 1  x
150 1    1 1  x
151 1    1 1  1
200 0    1 1  1
250 1    1 1  1
300 0    1 1  1
350 1    1 1  1
```

11.6.3   Wait Statement

The **wait statement** [Verilog LRM9.7.5] suspends a procedure until a condition becomes true. There must be another concurrent procedure that alters the condition (in this case the variable `Done`—in general the condition is an expression) in the following `wait` statement; otherwise we are placed on "infinite hold":

```
wait (Done) $stop; // Wait until Done = 1 then stop.
```

Notice that the Verilog `wait` statement does not look for an event or a change in the condition; instead it is level-sensitive—it only cares that the condition is true.

```
module test_dff_wait;                                                  //1
reg D, Clock, Reset; dff_wait u1(D, Q, Clock, Reset);                  //2
initial begin D=1; Clock=0;Reset=1'b1; #15 Reset=1'b0; #20 D=0; end    //3
always #10 Clock = !Clock;                                             //4
initial begin $display("T   Clk D Q Reset");                           //5
   $monitor("%2g",$time,,Clock,,,,D,,Q,,Reset); #50 $finish; end       //6
endmodule                                                              //7

module dff_wait(D, Q, Clock, Reset);                                   //1
output Q; input D, Clock, Reset; reg Q; wire D;                        //2
always @(posedge Clock) if (Reset !== 1) Q = D;                        //3
always begin wait (Reset == 1) Q = 0; wait (Reset !== 1); end          //4
endmodule                                                              //5
```

```
T  Clk D Q Reset
 0 0    1 0 1
10 1    1 0 1
15 1    1 0 0
20 0    1 0 0
```

```
30 1   1 1 0
35 1   0 1 0
40 0   0 1 0
```

We must include `wait` statements in module `dff_wait` above to wait for both `Reset==1` and `Reset==0`. If we were to omit the `wait` statement for `Reset==0`, as in the following code:

```
module dff_wait(D,Q,Clock,Reset);                              //1
output Q; input D,Clock,Reset; reg Q; wire D;                  //2
always @(posedge Clock) if (Reset !== 1) Q = D;                //3
// We need another wait statement here or we shall spin forever. //4
always begin wait (Reset == 1) Q = 0; end                      //5
endmodule                                                      //6
```

the simulator would cycle endlessly, and we would need to press the `'Stop'` button or `'CTRL-C'` to halt the simulator. Here is the console window in VeriWell:

```
C1> .
T  Clk D Q Reset      <- at this point nothing happens, so press CTRL-C
Interrupt at time 0
C1>
```

11.6.4 Blocking and Nonblocking Assignments

If a procedural assignment in a sequential block contains a timing control, then the execution of the following statement is delayed or **blocked**. For this reason a procedural assignment statement is also known as a **blocking procedural assignment statement** [Verilog LRM 9.2]. We covered this type of statement in Section 11.5.3. The **nonblocking procedural assignment statement** allows execution in a sequential block to continue and registers are all updated together at the end of the current time step. Both types of procedural assignment may contain timing controls. Here is an artificially complicated example that illustrates the different types of assignment:

```
module delay;                                                  //1
reg a,b,c,d,e,f,g,bds,bsd;                                     //2
initial begin                                                  //3
a = 1; b = 0;    // No delay control.                          //4
#1 b = 1;        // Delayed assignment.                        //5
c = #1 1;        // Intra-assignment delay.                    //6
#1;              // Delay control.                             //7
d = 1;           //                                            //8
e <= #1 1;       // Intra-assignment delay, nonblocking assignment //9
#1 f <= 1;       // Delayed nonblocking assignment.            //10
g <= 1;          // Nonblocking assignment.                    //11
end                                                            //12
initial begin #1 bds = b; end // Delay then sample (ds).       //13
initial begin bsd = #1 b; end // Sample then delay (sd).       //14
initial begin $display("t a b c d e f g bds bsd");             //15
```

```
$monitor("%g",$time,,a,,b,,c,,d,,e,,f,,g,,bds,,,,bsd); end    //16
endmodule                                                      //17
```

```
t a b c d e f g bds bsd
0 1 0 x x x x x x   x
1 1 1 x x x x x 1   0
2 1 1 1 x x x x 1   0
3 1 1 1 1 x x x 1   0
4 1 1 1 1 1 1 1 1   0
```

Many synthesis tools will not allow us to use blocking and nonblocking procedural assignments to the same reg within the same sequential block.

11.6.5   Procedural Continuous Assignment

A **procedural continuous assignment statement** [Verilog LRM 9.3] (sometimes called a quasicontinuous assignment statement) is a special form of the assign statement that we use within a sequential block. For example, the following flip-flop model assigns to q depending on the clear, clr_, and preset, pre_, inputs (in general it is considered very bad form to use a trailing underscore to signify active-low signals as I have done to save space; you might use "_n" instead).

```
module dff_procedural_assign;                                        //1
reg d,clr_,pre_,clk; wire q; dff_clr_pre dff_1(q,d,clr_,pre_,clk);   //2
always #10 clk = ~clk;                                               //3
initial begin clk = 0; clr_ = 1; pre_ = 1; d = 1;                    //4
  #20; d = 0; #20; pre_ = 0; #20; pre_ = 1; #20; clr_ = 0;           //5
  #20; clr_ = 1; #20; d = 1; #20; $finish; end                       //6
initial begin                                                        //7
  $display("T  CLK PRE_ CLR_ D Q");                                  //8
  $monitor("%3g",$time,,,clk,,,,pre_,,,,,clr_,,,,d,,q); end          //9
endmodule                                                            //10
```

```
module dff_clr_pre(q,d,clear_,preset_,clock);                //1
output q; input d,clear_,preset_,clock; reg q;               //2
always @(clear_ or preset_)                                  //3
  if (!clear_) assign q = 0; // active-low clear             //4
  else if(!preset_) assign q = 1; // active-low preset        //5
  else deassign q;                                           //6
always @(posedge clock) q = d;                               //7
endmodule                                                    //8
```

```
T  CLK PRE_ CLR_ D Q
 0  0   1    1   1 x
10  1   1    1   1 1
20  0   1    1   0 1
30  1   1    1   0 0
40  0   0    1   0 1
50  1   0    1   0 1
```

```
 60  0   1   1   0 1
 70  1   1   1   0 0
 80  0   1   0   0 0
 90  1   1   0   0 0
100  0   1   1   0 0
110  1   1   1   0 0
120  0   1   1   1 0
130  1   1   1   1 1
```

We have now seen all of the different forms of Verilog assignment statements. The following skeleton code shows where each type of statement belongs:

```
module all_assignments                                    //1
//... continuous assignments.                             //2
always // beginning of procedure                          //3
  begin // beginning of sequential block                  //4
  //... blocking procedural assignments.                  //5
  //... nonblocking procedural assignments.               //6
  //... procedural continuous assignments.                //7
  end                                                     //8
endmodule                                                 //9
```

Table 11.4 summarizes the different types of assignments.

TABLE 11.4   Verilog assignment statements.

Type of Verilog assignment	Continuous assignment statement	Procedural assignment statement	Nonblocking procedural assignment statement	Procedural continuous assignment statement
Where it can occur	outside an always or initial statement, task, or function	inside an always or initial statement, task, or function	inside an always or initial statement, task, or function	always or initial statement, task, or function
Example	wire [31:0] DataBus; assign DataBus = Enable ? Data : 32'bz	reg Y; always @(posedge clock) Y = 1;	reg Y; always Y <= 1;	always @(Enable) if(Enable) assign Q = D; else deassign Q;
Valid LHS of assignment	net	register or memory element	register or memory element	net
Valid RHS of assignment	<expression> net, reg or memory element	<expression> net, reg or memory element	<expression> net, reg or memory element	<expression> net, reg or memory element
Book	11.5.1	11.5.3	11.6.4	11.6.5
Verilog LRM	6.1	9.2	9.2.2	9.3

11.7   Tasks and Functions

A **task** [Verilog LRM 10.2] is a type of procedure, called from another procedure. A task has both inputs and outputs but does not return a value. A task may call other tasks and functions. A **function** [Verilog LRM 10.3] is a procedure used in any expression, has at least one input, no outputs, and returns a single value. A function may not call a task. In Section 11.5 we covered all of the different Verilog procedures except for tasks and functions. Now that we have covered timing controls, we can explain the difference between tasks and functions: Tasks may contain timing controls but functions may not. The following two statements help illustrate the difference between a function and a task:

```
Call_A_Task_And_Wait (Input1, Input2, Output);
Result_Immediate = Call_A_Function (All_Inputs);
```

Functions are useful to model combinational logic (rather like a subroutine):

```
module F_subset_decode; reg [2:0]A, B, C, D, E, F;              //1
initial begin A = 1; B = 0; D = 2; E = 3;                       //2
  C = subset_decode(A, B); F = subset_decode(D,E);              //3
  $display("A B C D E F"); $display(A,,B,,C,,D,,E,,F); end       //4
function [2:0] subset_decode; input [2:0] a, b;                 //5
  begin if (a <= b) subset_decode = a; else subset_decode = b; end //6
endfunction                                                     //7
endmodule                                                       //8

A B  C D  E F
1 0  0 2  3 2
```

11.8   Control Statements

In this section we shall discuss the Verilog if, case, loop, disable, fork, and join statements that control the flow of code execution.

11.8.1   Case and If Statement

An **if statement** [Verilog LRM 9.4] represents a two-way branch. In the following example, switch has to be true to execute 'Y = 1'; otherwise 'Y = 0' is executed:

```
if(switch) Y = 1; else Y = 0;
```

The **case statement** [Verilog LRM 9.5] represents a multiway branch. A **controlling expression** is matched with **case expressions** in each of the **case items** (or arms) to determine a match,

```
module test_mux; reg a, b, select; wire out;                   //1
mux mux_1(a, b, out, select);                                  //2
```

```
initial begin #2; select = 0; a = 0; b = 1;                  //3
   #2; select = 1'bx; #2; select = 1'bz; #2; select = 1; end //4
initial $monitor("T=%2g",$time," Select=",select," Out=",out); //5
initial #10 $finish;                                          //6
endmodule                                                     //7

module mux(a, b, mux_output, mux_select); input a, b, mux_select; //1
output mux_output; reg mux_output;                            //2
always begin                                                  //3
case(mux_select)                                              //4
   0: mux_output = a;                                         //5
   1: mux_output = b;                                         //6
   default mux_output = 1'bx; // If select = x or z set output to x. //7
endcase                                                       //8
#1; // Need some delay, otherwise we'll spin forever.         //9
end                                                           //10
endmodule                                                     //11
T= 0   Select=x   Out=x
T= 2   Select=0   Out=x
T= 3   Select=0   Out=0
T= 4   Select=x   Out=0
T= 5   Select=x   Out=x
T= 6   Select=z   Out=x
T= 8   Select=1   Out=x
T= 9   Select=1   Out=1
```

Notice that the case statement must be inside a sequential block (inside an always statement). Because the case statement is inside an always statement, it needs some delay; otherwise the simulation runs forever without advancing simulation time. The **casex statement** handles both 'z' and 'x' as don't care (so that they match any bit value), the **casez statement** handles 'z' bits, and only 'z' bits, as don't care. Bits in case expressions may be set to '?' representing don't care values, as follows:

```
casex (instruction_register[31:29])
   3b'??1 : add;
   3b'?1? : subtract;
   3b'1?? : branch;
endcase
```

11.8.2 Loop Statement

A **loop statement** [Verilog LRM 9.6] is a **for, while, repeat,** or **forever** statement. Here are four examples, one for each different type of loop statement, each of which performs the same function. The comments with each type of loop statement illustrate how the controls work:

```
module loop_1;                                               //1
integer i; reg [31:0] DataBus; initial DataBus = 0;          //2
```

```
initial begin                                                    //3
/************** Insert loop code after here. ******************/
/* for(Execute this assignment once before starting loop; exit loop if
this expression is false; execute this assignment at end of loop before
the check for end of loop.) */
for(i = 0; i <= 15; i = i+1) DataBus[i] = 1;                     //4
/************** Insert loop code before here. ***************/
end                                                              //5
initial begin                                                    //6
$display("DataBus = %b",DataBus);                                //7
#2; $display("DataBus = %b",DataBus); $finish;                   //8
end                                                              //9
endmodule                                                        //10
```

Here is the while statement code (to replace line 4 in module loop_1):

```
i = 0;
/* while(Execute next statement while this expression is true.) */
while(i <= 15) begin DataBus[i] = 1; i = i+1; end               //4
```

Here is the repeat statement code (to replace line 4 in module loop_1):

```
i = 0;
/* repeat(Execute next statement the number of times corresponding to
the evaluation of this expression at the beginning of the loop.) */
repeat(16) begin DataBus[i] = 1; i = i+1; end                   //4
```

Here is the forever statement code (to replace line 4 in module loop_1):

```
i = 0;
/* A forever statement loops continuously. */
forever begin : my_loop
  DataBus[i] = 1;
  if (i == 15) #1 disable my_loop; // Need to let time advance to exit.
  i = i+1;
end                                                              //4
```

The output for all four forms of looping statement is the same:

```
DataBus = 00000000000000000000000000000000
DataBus = 00000000000000001111111111111111
```

11.8.3 Disable

The **disable statement** [Verilog LRM 11] stops the execution of a labeled sequential block and skips to the end of the block:

```
forever
begin: microprocessor_block // Labeled sequential block.
  @(posedge clock)
  if (reset) disable microprocessor_block; // Skip to end of block.
```

```
    else Execute_code;
end
```

Use the `disable` statement with caution in ASIC design. It is difficult to implement directly in hardware.

11.8.4   Fork and Join

The **fork statement** and **join statement** [Verilog LRM 9.8.2] allows the execution of two or more parallel threads in a **parallel block**:

```
module fork_1                                     //1
event eat_breakfast, read_paper;                  //2
initial begin                                     //3
  fork                                            //4
  @eat_breakfast; @read_paper;                    //5
  join                                            //6
end                                               //7
endmodule                                         //8
```

This is another Verilog language feature that should be used with care in ASIC design, because it is difficult to implement in hardware.

11.9   Logic-Gate Modeling

Verilog has a set of built-in logic models and you may also define your own models.

11.9.1   Built-in Logic Models

Verilog's built-in logic models are the following **primitives** [Verilog LRM7]:

        and, nand, nor, or, xor, xnor

You may use these primitives as you use modules. For example:

```
module primitive;                                 //1
nand (strong0, strong1) #2.2                      //2
  Nand_1(n001, n004, n005),                       //3
  Nand_2(n003, n001, n005, n002);                 //4
nand (n006, n005, n002);                          //5
endmodule                                         //6
```

This module models three NAND gates (Figure 11.2). The first gate (line 3) is a two-input gate named `Nand_1`; the second gate (line 4) is a three-input gate named `Nand_2`; the third gate (line 5) is unnamed. The first two gates have strong drive strengths [Verilog LRM3.4] (these are the defaults anyway) and 2.2 ns delay; the third gate takes the default values for drive strength (strong) and delay (zero). The first port of a primitive gate is always the output port. The remaining ports for a primitive gate (any number of them) are the input ports.

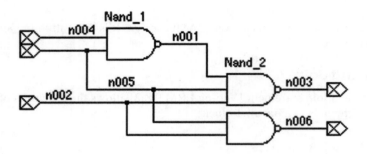

FIGURE 11.2 An example schematic (drawn with Capilano's DesignWorks) to illustrate the use of Verilog primitive gates.

Table 11.5 shows the definition of the and gate primitive (I use lowercase 'and' as the name of the Verilog primitive, rather than 'AND', since Verilog is case-sensitive). Notice that if one input to the primitive 'and' gate is zero, the output is zero, no matter what the other input is.

TABLE 11.5   Definition of the Verilog primitive 'and' gate.

'and'	0	1	x	z
0	0	0	0	0
1	0	1	x	x
x	0	x	x	x
z	0	x	x	x

11.9.2   User-Defined Primitives

We can define primitive gates (a **user-defined primitive** or **UDP**) using a truth-table specification [Verilog LRM8]. The first port of a UDP must be an output port, and this must be the only output port (we may not use vector or inout ports):

```
primitive Adder(Sum, InA, InB);                                          //1
output Sum; input Ina, InB;                                              //2
table                                                                    //3
// inputs : output                                                       //4
00 : 0;                                                                  //5
01 : 1;                                                                  //6
10 : 1;                                                                  //7
```

```
11 : 0;                                                            //8
endtable                                                           //9
endprimitive                                                       //10
```

We may only specify the values `'0'`, `'1'`, and `'x'` as inputs in a **UDP truth table**. Any `'z'` input is treated as an `'x'`. If there is no entry in a UDP truth table that exactly matches a set of inputs, the output is `'x'` (unknown).

We can construct a UDP model for sequential logic by including a state in the UDP truth-table definition. The state goes between an input and an output in the table and the output then represents the next state. The following sequential UDP model also illustrates the use of shorthand notation in a UDP truth table:

```
primitive DLatch(Q, Clock, Data);                                  //1
output Q; reg Q; input Clock, Data;                                //2
table                                                              //3
//inputs : present state : output (next state)                     //4
1 0 : ? : 0; // ? represents 0,1, or x (input or present state).   //5
1 1 : b : 1; // b represents 0 or 1 (input or present state).      //6
1 1 : x : 1; // Could have combined this with previous line.       //7
0 ? : ? : -; // - represents no change in an output.               //8
endtable                                                           //9
endprimitive                                                       //10
```

Be careful not to confuse the `'?'` in a UDP table (shorthand for `'0'`, `'1'`, or `'x'`) with the `'?'` in a constant that represents an extension to `'z'` (Section 11.2.4) or the `'?'` in a `case` statement that represents don't care values (Section 11.8.1).

For sequential UDP models that need to detect edge transitions on inputs, there is another special truth-table notation `(ab)` that represents a change in logic value from a to b. For example, `(01)` represents a rising edge. There are also shorthand notations for various edges:

- `*` is `(??)`
- `r` is `(01)`
- `f` is `(10)`
- `p` is `(01)`, `(0x)`, or `(x1)`
- `n` is `(10)`, `(1x)`, or `(x0)`

```
primitive DFlipFlop(Q, Clock, Data);                               //1
output Q; reg Q; input Clock, Data;                                //2
table                                                              //3
//inputs : present state : output (next state)                     //4
r    0 : ? : 0 ; // rising edge, next state = output = 0           //5
r    1 : ? : 1 ; // rising edge, next state = output = 1           //6
(0x) 0 : 0 : 0 ; // rising edge, next state = output = 0           //7
(0x) 1 : 1 : 1 ; // rising edge, next state = output = 1           //8
(?0) ? : ? : - ; // falling edge, no change in output             //9
? (??) : ? : - ; // no clock edge, no change in output            //10
```

```
endtable                                                    //11
endprimitive                                                //12
```

11.10 Modeling Delay

Verilog has a set of built-in methods to define delays. This is very important in ASIC physical design. Before we start layout, we can use ASIC cell library models written in Verilog that include logic delays as a function of fanout and estimated wiring loads. After we have completed layout, we can extract the wiring capacitance, allowing us to calculate the exact delay values. Using the techniques described in this section, we can then back-annotate our Verilog netlist with postlayout delays and complete a postlayout simulation.

We can complete this back-annotation process in a standard fashion since delay specification is part of the Verilog language. This makes working with an ASIC cell library and the ASIC foundry that will fabricate our ASIC much easier. Typically an ASIC library company might sell us a cell library complete with Verilog models that include all the minimum, typical, and maximum delays as well as the different values for rising and falling transitions. The ASIC foundry will provide us with a delay calculator that calculates the net delays (this is usually proprietary technology) from the layout. These delays are held in a separate file (the **Standard Delay Format**, **SDF**, is widely used) and then mapped to parameters in the Verilog models. If we complete back-annotation and a postlayout simulation using an approved cell library, the ASIC foundry will "sign off" on our design. This is basically a guarantee that our chip will work according to the simulation. This ability to design sign-off quality ASIC cell libraries is very important in the ASIC design process.

11.10.1 Net and Gate Delay

We saw how to specify a delay control for any statement in Section 11.6. In fact, Verilog allows us to specify minimum, typical, and maximum values for the delay as follows [Verilog LRM7.15]:

```
#(1.1:1.3:1.7) assign delay_a = a; // min:typ:max
```

We can also specify the delay properties of a `wire` in a similar fashion:

```
wire #(1.1:1.3:1.7) a_delay; // min:typ:max
```

We can specify delay in a `wire` declaration together with a continuous assignment as in the following example:

```
wire #(1.1:1.3:1.7) a_delay = a; // min:typ:max
```

but in this case the delay is associated with the driver and not with the `wire`.

In Section 11.9.1 we explained that we can specify a delay for a logic primitive. We can also specify minimum, typical, and maximum delays as well as separate delays for rising and falling transitions for primitives as follows [Verilog LRM4.3]:

```
nand #3.0 nd01(c, a, b);
nand #(2.6:3.0:3.4) nd02(d, a, b); // min:typ:max
nand #(2.8:3.2:3.4, 2.6:2.8:2.9) nd03(e, a, b);
// #(rising, falling) delay
```

The first NAND gate, nd01, has a delay of 3 ns (assuming we specified nanoseconds as the timescale) for both rising and falling delays. The NAND gate nd02 has a triplet for the delay; this corresponds to a minimum (2.6 ns), typical (3.0 ns), and a maximum delay (3.4 ns). The NAND gate nd03 has two triplets for the delay: The first triplet specifies the min/typ/max rising delay ('0' or 'x' or 'z' to '1'), and the second triplet specifies the min/typ/max falling delay ('1' or 'x' or 'z' to '0').

Some primitives can produce a high-impedance output, 'z'. In this case we can specify a triplet of delay values corresponding to rising transition, falling transition, and the delay to transition to 'z' (from '0' or '1' to 'z'—this is usually the delay for a three-state driver to turn off or float). We can do the same thing for net types,

```
wire #(0.5,0.6,0.7) a_z = a; // rise/fall/float delays
```

11.10.2 Pin-to-Pin Delay

The **specify block** [Verilog LRM 13] is a special construct in Verilog that allows the definition of **pin-to-pin delays** across a module. The use of a specify block can include the use of built-in system functions to check setup and hold times, for example. The following example illustrates how to specify pin-to-pin timing for a D flip-flop. We declare the timing parameters first followed by the paths. This example uses the UDP from Section 11.9.2, which does not include preset and clear (so only part of the flip-flop function is modeled), but includes the timing for preset and clear for illustration purposes.

```
module DFF_Spec; reg D, clk;                              //1
DFF_Part DFF1 (Q, clk, D, pre, clr);                      //2
initial begin D = 0; clk = 0; #1; clk = 1; end            //3
initial $monitor("T=%2g", $time," clk=", clk," Q=", Q);   //4
endmodule                                                 //5

module DFF_Part(Q, clk, D, pre, clr);                     //1
   input clk, D, pre, clr; output Q;                      //2
   DFlipFlop(Q, clk, D); // No preset or clear in this UDP. //3
   specify                                                //4
      specparam                                           //5
      tPLH_clk_Q = 3, tPHL_clk_Q = 2.9,                   //6
      tPLH_set_Q = 1.2, tPHL_set_Q = 1.1;                 //7
   (clk => Q) = (tPLH_clk_Q, tPHL_clk_Q);                 //8
   (pre, clr *> Q) = (tPLH_set_Q, tPHL_set_Q);            //9
```

```
    endspecify                                                          //10
    endmodule                                                           //11

T= 0 clk=0 Q=x
T= 1 clk=1 Q=x
T= 4 clk=1 Q=0
```

There are the following two ways to specify paths (module `DFF_part` above uses both) [Verilog LRM13.3]:

- `x => y` specifies a **parallel connection** (or parallel path) between x and y (x and y must have the same number of bits).

- `x *> y` specifies a **full connection** (or full path) between x and y (every bit in x is connected to y). In this case x and y may be different sizes.

The delay of some logic cells depends on the state of the inputs. This can be modeled using a **state-dependent path delay**. Here is an example:

```
`timescale 1 ns / 100 fs                                                //1
module M_Spec; reg A1, A2, B; M M1 (Z, A1, A2, B);                      //2
initial begin A1=0;A2=1;B=1;#5;B=0;#5;A1=1;A2=0;B=1;#5;B=0; end          //3
initial                                                                 //4
  $monitor("T=%4g",$realtime," A1=",A1," A2=",A2," B=",B," Z=",Z);       //5
endmodule                                                               //6

`timescale 100 ps / 10 fs                                               //1
module M(Z, A1, A2, B); input A1, A2, B; output Z;                      //2
or (Z1, A1, A2); nand (Z, Z1, B); // OAI21                              //3
/*A1 A2 B Z   Delay=10*100 ps unless indicated in the table below.      //4
  0  0  0 1                                                             //5
  0  0  1 1                                                             //6
  0  1  0 1  B:0->1 Z:1->0 delay=t2                                     //7
  0  1  1 0  B:1->0 Z:0->1 delay=t1                                     //8
  1  0  0 1  B:0->1 Z:1->0 delay=t4                                     //9
  1  0  1 0  B:1->0 Z:0->1 delay=t3                                     //10
  1  1  0 1                                                             //11
  1  1  1 0 */                                                          //12
specify specparam t1 = 11, t2 = 12; specparam t3 = 13, t4 = 14;         //13
  (A1 => Z) = 10; (A2 => Z) = 10;                                       //14
  if (~A1) (B => Z) = (t1, t2); if (A1) (B => Z) = (t3, t4);            //15
endspecify                                                              //16
endmodule                                                               //17

T=   0 A1=0 A2=1 B=1 Z=x
T=   1 A1=0 A2=1 B=1 Z=0
T=   5 A1=0 A2=1 B=0 Z=0
T= 6.1 A1=0 A2=1 B=0 Z=1
T=  10 A1=1 A2=0 B=1 Z=1
T=  11 A1=1 A2=0 B=1 Z=0
T=  15 A1=1 A2=0 B=0 Z=0
T=16.3 A1=1 A2=0 B=0 Z=1
```

11.11 Altering Parameters

Here is an example of a module that uses a parameter [Verilog LRM3.10, 12.2]:

```
module Vector_And(Z, A, B);                                  //1
  parameter CARDINALITY = 1;                                 //2
  input [CARDINALITY-1:0] A, B;                              //3
  output [CARDINALITY-1:0] Z;                                //4
  wire [CARDINALITY-1:0] Z = A & B;                          //5
endmodule                                                    //6
```

We can override this parameter when we instantiate the module as follows:

```
module Four_And_Gates(OutBus, InBusA, InBusB);               //1
  input [3:0] InBusA, InBusB; output [3:0] OutBus;           //2
  Vector_And #(4) My_AND(OutBus, InBusA, InBusB); // 4 AND gates  //3
endmodule                                                    //4
```

The parameters of a module have local scope, but we may override them using a **defparam** statement and a hierarchical name, as in the following example:

```
module And_Gates(OutBus, InBusA, InBusB);                    //1
  parameter WIDTH = 1;                                       //2
  input [WIDTH-1:0] InBusA, InBusB; output [WIDTH-1:0] OutBus;  //3
  Vector_And #(WIDTH) My_And(OutBus, InBusA, InBusB);        //4
endmodule                                                    //5

module Super_Size; defparam And_Gates.WIDTH = 4; endmodule   //1
```

11.12 A Viterbi Decoder

This section describes an ASIC design for a Viterbi decoder using Verilog. Christeen Gray completed the original design as her MS thesis at the University of Hawaii (UH) working with VLSI Technology, using the Compass ASIC Synthesizer and a VLSI Technology cell library. The design was mapped from VLSI Technology design rules to Hewlett-Packard design rules; prototypes were fabricated by Hewlett-Packard (through Mosis) and tested at UH.

11.12.1 Viterbi Encoder

Viterbi encoding is widely used for satellite and other noisy **communications channels**. There are two important components of a channel using Viterbi encoding: the **Viterbi encoder** (at the transmitter) and the **Viterbi decoder** (at the receiver). A

Viterbi encoder includes extra information in the transmitted signal to reduce the probability of errors in the received signal that may be corrupted by noise.

I shall describe an encoder in which every two bits of a data stream are encoded into three bits for transmission. The ratio of input to output information in an encoder is the **rate** of the encoder; this is a rate 2/3 encoder. The following equations relate the three encoder output bits (Y_n^2, Y_n^1, and Y_n^0) to the two encoder input bits (X_n^2 and X_n^1) at a time nT:

$$Y_n^2 = X_n^2$$
$$Y_n^1 = X_n^1 \oplus X_{n-2}^1 \tag{11.1}$$
$$Y_n^0 = X_{n-1}^1$$

We can write the input bits as a single number. Thus, for example, if $X_n^2 = 1$ and $X_n^1 = 0$, we can write $X_n = 2$. Equation 11.1 defines a state machine with two memory elements for the two last input values for X_n^1: X_{n-1}^1 and X_{n-2}^1. These two state variables define four states: $\{X_{n-1}^1, X_{n-2}^1\}$, with $S_0 = \{0, 0\}$, $S_1 = \{1, 0\}$, $S_2 = \{0, 1\}$, and $S_3 = \{1, 1\}$. The 3-bit output Y_n is a function of the state and current 2-bit input X_n.

The following Verilog code describes the rate 2/3 encoder. This model uses two D flip-flops as the state register. When reset (using active-high input signal **res**) the encoder starts in state S_0. In Verilog I represent Y_n^2 by **Y2N**, for example.

```
/***********************************************************/
/* module viterbi_encode                                   */
/***********************************************************/
/* This is the encoder. X2N (msb) and X1N form the 2-bit input
message, XN. Example: if X2N=1, X1N=0, then XN=2. Y2N (msb), Y1N, and
Y0N form the 3-bit encoded signal, YN (for a total constellation of 8
PSK signals that will be transmitted). The encoder uses a state
machine with four states to generate the 3-bit output, YN, from the
2-bit input, XN. Example: the repeated input sequence XN = (X2N, X1N)
= 0, 1, 2, 3 produces the repeated output sequence YN = (Y2N, Y1N,
Y0N) = 1, 0, 5, 4. */
module viterbi_encode(X2N,X1N,Y2N,Y1N,Y0N,clk,res);
input X2N,X1N,clk,res; output Y2N,Y1N,Y0N;
wire X1N_1,X1N_2,Y2N,Y1N,Y0N;
dff dff_1(X1N,X1N_1,clk,res); dff dff_2(X1N_1,X1N_2,clk,res);
assign Y2N=X2N; assign Y1N=X1N ^ X1N_2; assign Y0N=X1N_1;
endmodule
```

Figure 11.3 shows the state diagram for this encoder. The first four rows of Table 11.6 show the four different transitions that can be made from state S_0. For example, if we reset the encoder and the input is $X_n = 3$ ($X_n^2 = 1$ and $X_n^1 = 1$), then the output will be $Y_n = 6$ ($Y_n^2 = 1$, $Y_n^1 = 1$, $Y_n^0 = 0$) and the next state will be S_1.

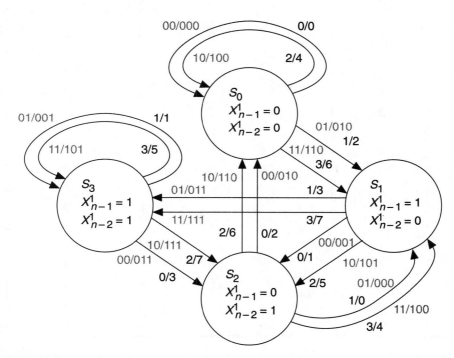

FIGURE 11.3 A state diagram for a rate 2/3 Viterbi encoder. The inputs and outputs are shown in binary as $X_n^2 X_n^1 / Y_n^2 Y_n^1 Y_n^0$, and in decimal as X_n / Y_n.

As an example, the repeated encoder input sequence $X_n = 0, 1, 2, 3, \ldots$ produces the encoder output sequence $Y_n = 1, 0, 5, 4, \ldots$ repeated. Table 11.7 shows the state transitions for this sequence, including the initialization steps.

Next we transmit the eight possible encoder outputs ($Y_n = 0\text{--}7$) as **signals** over our noisy communications channel (perhaps a microwave signal to a satellite) using the **signal constellation** shown in Figure 11.4. Typically this is done using **phase-shift keying (PSK)** with each signal position corresponding to a different phase shift in the transmitted carrier signal.

TABLE 11.6 State table for the rate 2/3 Viterbi encoder.

Present state	Inputs		State variables		Y_n^2 $= X_n^2$	Y_n^1 $= X_n^1 \oplus X_{n-2}^1$	Y_n^0 $= X_{n-1}^1$	Next state $\{X_{n-1}^1, X_{n-2}^1\}$	
	X_n^2	X_n^1	X_{n-1}^1	X_{n-2}^1					
S_0	0	0	0	0	0	0	0	00	S_0
S_0	0	1	0	0	0	1	0	10	S_1
S_0	1	0	0	0	1	0	0	00	S_0
S_0	1	1	0	0	1	1	0	10	S_1
S_1	0	0	1	0	0	0	1	01	S_2
S_1	0	1	1	0	0	1	1	11	S_3
S_1	1	0	1	0	1	0	1	01	S_2
S_1	1	1	1	0	1	1	1	11	S_3
S_2	0	0	0	1	0	1	0	00	S_0
S_2	0	1	0	1	0	0	0	10	S_1
S_2	1	0	0	1	1	1	0	00	S_0
S_2	1	1	0	1	1	0	0	10	S_1
S_3	0	0	1	1	0	1	1	01	S_2
S_3	0	1	1	1	0	0	1	11	S_3
S_3	1	0	1	1	1	1	1	01	S_2
S_3	1	1	1	1	1	0	1	11	S_3

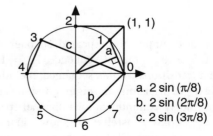

FIGURE 11.4 The signal constellation for an 8PSK (phase-shift keyed) code.

a. $2 \sin (\pi/8)$
b. $2 \sin (2\pi/8)$
c. $2 \sin (3\pi/8)$

TABLE 11.7   A sequence of transmitted signals for the rate 2/3 Viterbi encoder

Time ns	Inputs		State variables		Outputs			Present state	Next state
	X_n^2	X_n^1	X_{n-1}^1	X_{n-2}^1	Y_n^2	Y_n^1	Y_n^0		
0	1	1	x	x	1	x	x	$S_?$	$S_?$
10	1	1	0	0	1	1	0	S_0	S_1
50	0	0	1	0	0	0	1	S_1	S_2
150	0	1	0	1	0	0	0	S_2	S_1
250	1	0	1	0	1	0	1	S_1	S_2
350	1	1	0	1	1	0	0	S_2	S_1
450	0	0	1	0	0	0	1	S_1	S_2
550	0	1	0	1	0	0	0	S_2	S_1
650	1	0	1	0	1	0	1	S_1	S_2
750	1	1	0	1	1	0	0	S_2	S_1
850	0	0	1	0	0	0	1	S_1	S_2
950	0	1	0	1	0	0	0	S_2	S_1

11.12.2 The Received Signal

The noisy signal enters the receiver. It is now our task to discover which of the eight possible signals were transmitted at each time step. First we calculate the distance of each received signal from each of the known eight positions in the signal constellation. Table 11.8 shows the distances between signals in the 8PSK constellation. We are going to assume that there is no noise in the channel to illustrate the operation of the Viterbi decoder, so that the distances in Table 11.8 represent the possible distance measures of our received signal from the 8PSK signals.

The distances, X, in the first column of Table 11.8 are the geometric or algebraic distances. We measure the **Euclidean distance**, $E = X^2$ shown as B (the binary quantized value of E) in Table 11.8. The rounding errors that result from conversion to fixed-width binary are **quantization errors** and are important in any practical implementation of the Viterbi decoder. The effect of the quantization error is to add a form of noise to the received signal.

The following code models the receiver section that digitizes the noisy analog received signal and computes the binary distance measures. Eight binary-distance measures, in0-in7, are generated each time a signal is received. Since each of the distance measures is 3 bits wide, there are a total of 24 bits (8×3) that form the digital inputs to the Viterbi decoder.

TABLE 11.8   Distance measures for Viterbi encoding (8PSK).

Signal	Algebraic distance from signal 0	X=Distance from signal 0	Euclidean distance $E = X^2$	B = binary quantized value of E	D=decimal value of B	Quantization error $Q = D - 1.75E$
0	$2\sin(0\pi/8)$	0.00	0.00	000	0	0
1	$2\sin(1\pi/8)$	0.77	0.59	001	1	−0.0325
2	$2\sin(2\pi/8)$	1.41	2.00	100	4	0.5
3	$2\sin(3\pi/8)$	1.85	3.41	110	6	0.0325
4	$2\sin(4\pi/8)$	2.00	4.00	111	7	0
5	$2\sin(5\pi/8)$	1.85	3.41	110	6	0.0325
6	$2\sin(6\pi/8)$	1.41	2.00	100	4	0.5
7	$2\sin(7\pi/8)$	0.77	0.59	001	1	−0.0325

```
/********************************************************/
/* module viterbi_distances                            */
/********************************************************/
/* This module simulates the front end of a receiver. Normally the
received analog signal (with noise) is converted into a series of
distance measures from the known eight possible transmitted PSK
signals: s0,...,s7. We are not simulating the analog part or noise in
this version, so we just take the digitally encoded 3-bit signal, Y,
from the encoder and convert it directly to the distance measures.
d[N] is the distance from signal = N to signal = 0
d[N] = (2*sin(N*PI/8))**2 in 3-bit binary (on the scale 2=100)
Example: d[3] = 1.85**2 = 3.41 = 110
inN is the distance from signal = N to encoder signal.
Example: in3 is the distance from signal = 3 to encoder signal.
d[N] is the distance from signal = N to encoder signal = 0.
If encoder signal = J, shift the distances by 8-J positions.
Example: if signal = 2, in0 is d[6], in1 is D[7], in2 is D[0], etc. */
module viterbi_distances
    (Y2N,Y1N,Y0N,clk,res,in0,in1,in2,in3,in4,in5,in6,in7);
input clk,res,Y2N,Y1N,Y0N; output in0,in1,in2,in3,in4,in5,in6,in7;
reg [2:0] J,in0,in1,in2,in3,in4,in5,in6,in7; reg [2:0] d [7:0];
initial begin d[0]='b000;d[1]='b001;d[2]='b100;d[3]='b110;
d[4]='b111;d[5]='b110;d[6]='b100;d[7]='b001; end
always @(Y2N or Y1N or Y0N) begin
J[0]=Y0N;J[1]=Y1N;J[2]=Y2N;
J=8-J;in0=d[J];J=J+1;in1=d[J];J=J+1;in2=d[J];J=J+1;in3=d[J];
J=J+1;in4=d[J];J=J+1;in5=d[J];J=J+1;in6=d[J];J=J+1;in7=d[J];
end endmodule
```

As an example, Table 11.9 shows the distance measures for the transmitted encoder output sequence Y_n = 1, 0, 5, 4, ... (repeated) corresponding to an encoder input of X_n = 0, 1, 2, 3, ... (repeated).

TABLE 11.9   Receiver distance measures for an example transmission sequence.

Time ns	Input X_n	Output Y_n	Present state	Next state	in0	in1	in2	in3	in4	in5	in6	in7
0	3	x	$S_?$	$S_?$	x	x	x	x	x	x	x	x
10	3	6	S_0	S_1	4	6	7	6	4	1	0	1
50	0	1	S_1	S_2	1	0	1	4	6	7	6	4
150	1	0	S_2	S_1	0	1	4	6	7	6	4	1
250	2	5	S_1	S_2	6	7	6	4	1	0	1	4
350	3	4	S_2	S_1	7	6	4	1	0	1	4	6
450	0	1	S_1	S_2	1	0	1	4	6	7	6	4
550	1	0	S_2	S_1	0	1	4	6	7	6	4	1
650	2	5	S_1	S_2	6	7	6	4	1	0	1	4
750	3	4	S_2	S_1	7	6	4	1	0	1	4	6
850	0	1	S_1	S_2	1	0	1	4	6	7	6	4
950	1	0	S_2	S_1	0	1	4	6	7	6	4	1

11.12.3   Testing the System

Here is a testbench for the entire system: encoder, receiver front end, and decoder:

```
/*******************************************************/
/* module viterbi_test_CDD                            */
/*******************************************************/
/* This is the top-level module, viterbi_test_CDD, that models the
communications link. It contains three modules: viterbi_encode,
viterbi_distances, and viterbi. There is no analog and no noise in
this version. The 2-bit message, X, is encoded to a 3-bit signal, Y.
In this module the message X is generated using a simple counter.
The digital 3-bit signal Y is transmitted, received with noise as an
analog signal (not modeled here), and converted to a set of eight
3-bit distance measures, in0, ..., in7. The distance measures form
the input to the Viterbi decoder that reconstructs the transmitted
signal Y, with an error signal if the measures are inconsistent.
CDD = counter input, digital transmission, digital reception */
module viterbi_test_CDD;
```

```
wire Error;                    // decoder out
wire [2:0] Y, Out;             // encoder out, decoder out
reg [1:0] X;                   // encoder inputs
reg Clk, Res;                  // clock and reset
wire [2:0] in0,in1,in2,in3,in4,in5,in6,in7;
always #500 $display("t    Clk X Y Out Error");
initial $monitor("%4g",$time,,Clk,,,,X,,Y,,Out,,,,Error);
initial $dumpvars; initial #3000 $finish;
always #50 Clk = ~Clk; initial begin Clk = 0;
X = 3;                         // No special reason to start at 3.
#60 Res = 1;#10 Res = 0;end    // Hit reset after inputs are stable.
always @(posedge Clk) #1 X = X + 1; // Drive the input with a counter.
viterbi_encode v_1
   (X[1],X[0],Y[2],Y[1],Y[0],Clk,Res);
viterbi_distances v_2
   (Y[2],Y[1],Y[0],Clk,Res,in0,in1,in2,in3,in4,in5,in6,in7);
viterbi v_3
   (in0,in1,in2,in3,in4,in5,in6,in7,Out,Clk,Res,Error);
endmodule
```

The Viterbi decoder takes the distance measures and calculates the most likely transmitted signal. It does this by keeping a running history of the previously received signals in a **path memory**. The path-memory length of this decoder is 12. By keeping a history of possible sequences and using the knowledge that the signals were generated by a state machine, it is possible to select the most likely sequences.

TABLE 11.10   Output from the Viterbi testbench

t	Clk	X	Y	Out	Error	t	Clk	X	Y	Out	Error
0	0	3	x	x	0	1351	1	1	0	0	0
50	1	3	x	x	0	1400	0	1	0	0	0
51	1	0	x	x	0	1450	1	1	0	0	0
60	1	0	0	0	0	1451	1	2	5	2	0
100	0	0	0	0	0	1500	0	2	5	2	0
150	1	0	0	0	0	1550	1	2	5	2	0
151	1	1	2	0	0	1551	1	3	4	5	0

Table 11.10 shows part of the simulation results from the testbench, viterbi_test_CDD, in tabular form. Figure 11.5 shows the Verilog simulator output from the testbench (displayed using VeriWell from Wellspring).

The system input or message, X[1:0], is driven by a counter that repeats the sequence 0, 1, 2, 3, ... incrementing by 1 at each positive clock edge (with a delay of one time unit), starting with X equal to 3 at $t = 0$. The active-high reset signal, Res, is asserted at $t = 60$ for 10 time units. The encoder output, Y[2:0], changes at $t = 151$, which is one time unit (the positive-edge–triggered D flip-flop model contains a

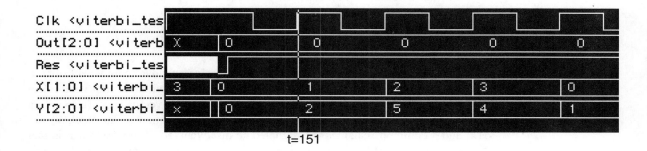

t=151

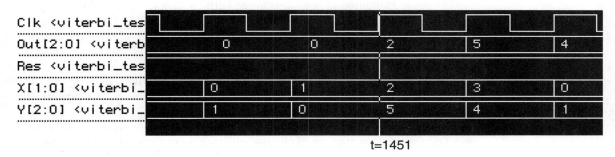

t=1451

FIGURE 11.5 Viterbi encoder testbench simulation results. (Top) Initialization and the start of the encoder output sequence 2, 5, 4, 1, 0, ... on Y[2:0] at *t* = 151. (Bottom) The appearance of the same encoder output sequence at the output of the decoder, Out[2:0], at *t* = 1451, 1300 time units (13 positive clock edges) later.

one-time-unit delay) after the first positive clock edge (at *t* = 150) following the deassertion of the reset at *t* = 70. The encoder output sequence beginning at *t* = 151 is 2, 5, 4, 1, 0, ... and then the sequence 5, 4, 1, 0, ... repeats. This encoder output sequence is then imagined to be transmitted and received. The receiver module calculates the distance measures and passes them to the decoder. After 13 positive clock-edges (1300 time ticks) the transmitted sequence appears at the output, Out[2:0], beginning at *t* = 1451 with 2, 5, 4, 1, 0, ..., exactly the same as the encoder output.

11.12.4 Verilog Decoder Model

The Viterbi decoder model presented in this section is written for both simulation and synthesis. The Viterbi decoder makes extensive use of vector D flip-flops (registers). Early versions of Verilog-XL did not support vector instantiations of modules. In addition the inputs of UDPs may not be vectors and there are no primitive D flip-flops in Verilog. This makes instantiation of a register difficult other than by writing a separate module instance for each flip-flop.

The first solution to this problem is to use flip-flop models supplied with the synthesis tool such as the following:

```
· asDff #(3) subout0(in0, sub0, clk, reset);
```

The asDff is a model in the Compass ASIC Synthesizer standard component library. This statement triggers the synthesis of three D flip-flops, with an input vector ina (with a range of three) connected to the D inputs, an output vector sub0 (also with a range of three) connected to the Q flip-flop outputs, a common scalar clock signal, clk, and a common scalar reset signal. The disadvantage of this approach is that the names, functional behavior, and interfaces of the standard components are different for every software system.

The second solution, in new versions of Verilog-XL and other tools that support the IEEE standard, is to use vector instantiation as follows [LRM 7.5.1, 12.1.2]:

```
myDff subout0[0:2] (in0, sub0, clk, reset);
```

This instantiates three copies of a user-defined module or UDP called myDff. The disadvantage of this approach is that not all simulators and synthesizers support vector instantiation.

The third solution (which is used in the Viterbi decoder model) is to write a model that supports vector inputs and outputs. Here is an example D flip-flop model:

```
/******************************************************/
/*          module dff                                */
/******************************************************/
/* A D flip-flop module. */
module dff(D,Q,Clock,Reset); // N.B. reset is active-low.
output Q; input D,Clock,Reset;
parameter CARDINALITY = 1; reg [CARDINALITY-1:0] Q;
wire [CARDINALITY-1:0] D;
always @(posedge Clock) if (Reset !== 0) #1 Q = D;
always begin wait (Reset == 0); Q = 0; wait (Reset == 1); end
endmodule
```

We use this model by defining a parameter that specifies the bus width as follows:

```
    dff #(3) subout0(in0, sub0, clk, reset);
```

The code that models the entire Viterbi decoder is listed below (Figure 12.6 on page 578 shows the block diagram). Notice the following:

- Comments explain the function of each module.
- Each module is about a page or less of code.

- Each module can be tested by itself.
- The code is as simple as possible avoiding clever coding techniques.

The code is not flexible, because bit widths are fixed rather than using parameters. A model with parameters for rate, signal constellation, distance measure resolution, and path memory length is considerably more complex. We shall use this Viterbi decoder design again when we discuss logic synthesis in Chapter 12, test in Chapter 14, floorplanning and placement in Chapter 16, and routing in Chapter 17.

```verilog
/* Verilog code for a Viterbi decoder. The decoder assumes a rate
2/3 encoder, 8 PSK modulation, and trellis coding. The viterbi module
contains eight submodules: subset_decode, metric, compute_metric,
compare_select, reduce, pathin, path_memory, and output_decision.
   The decoder accepts eight 3-bit measures of ||r-si||**2 and, after
an initial delay of thirteen clock cycles, the output is the best
estimate of the signal transmitted. The distance measures are the
Euclidean distances between the received signal r (with noise) and
each of the (in this case eight) possible transmitted signals s0 to s7.
   Original by Christeen Gray, University of Hawaii. Heavily modified
by MJSS; any errors are mine. Use freely. */
/***************************************************/
/*    module viterbi                               */
/***************************************************/
/* This is the top level of the Viterbi decoder. The eight input
signals {in0,...,in7} represent the distance measures, ||r-si||**2.
The other input signals are clk and reset. The output signals are
out and error. */
module viterbi
    (in0,in1,in2,in3,in4,in5,in6,in7,
    out,clk,reset,error);
input [2:0] in0,in1,in2,in3,in4,in5,in6,in7;
output [2:0] out; input clk,reset; output error;
wire sout0,sout1,sout2,sout3;
wire [2:0] s0,s1,s2,s3;
wire [4:0] m_in0,m_in1,m_in2,m_in3;
wire [4:0] m_out0,m_out1,m_out2,m_out3;
wire [4:0] p0_0,p2_0,p0_1,p2_1,p1_2,p3_2,p1_3,p3_3;
wire ACS0,ACS1,ACS2,ACS3;
wire [4:0] out0,out1,out2,out3;
wire [1:0] control;
wire [2:0] p0,p1,p2,p3;
wire [11:0] path0;

   subset_decode u1(in0,in1,in2,in3,in4,in5,in6,in7,
      s0,s1,s2,s3,sout0,sout1,sout2,sout3,clk,reset);
   metric u2(m_in0,m_in1,m_in2,m_in3,m_out0,
      m_out1,m_out2,m_out3,clk,reset);
   compute_metric u3(m_out0,m_out1,m_out2,m_out3,s0,s1,s2,s3,
```

```
          p0_0,p2_0,p0_1,p2_1,p1_2,p3_2,p1_3,p3_3,error);
       compare_select u4(p0_0,p2_0,p0_1,p2_1,p1_2,p3_2,p1_3,p3_3,
          out0,out1,out2,out3,ACS0,ACS1,ACS2,ACS3);
       reduce u5(out0,out1,out2,out3,
          m_in0,m_in1,m_in2,m_in3,control);
       pathin u6(sout0,sout1,sout2,sout3,
          ACS0,ACS1,ACS2,ACS3,path0,clk,reset);
       path_memory u7(p0,p1,p2,p3,path0,clk,reset,
          ACS0,ACS1,ACS2,ACS3);
       output_decision u8(p0,p1,p2,p3,control,out);
endmodule


/******************************************************/
/* module subset_decode                               */
/******************************************************/
/* This module chooses the signal corresponding to the smallest of
each set {||r-s0||**2,||r-s4||**2}, {||r-s1||**2, ||r-s5||**2},
{||r-s2||**2,||r-s6||**2}, {||r-s3||**2,||r-s7||**2}. Therefore
there are eight input signals and four output signals for the
distance measures. The signals sout0, ..., sout3 are used to control
the path memory. The statement dff #(3) instantiates a vector array
of 3 D flip-flops. */
module subset_decode
       (in0,in1,in2,in3,in4,in5,in6,in7,
       s0,s1,s2,s3,
       sout0,sout1,sout2,sout3,
       clk,reset);
input [2:0] in0,in1,in2,in3,in4,in5,in6,in7;
output [2:0] s0,s1,s2,s3;
output sout0,sout1,sout2,sout3;
input clk,reset;
wire [2:0] sub0,sub1,sub2,sub3,sub4,sub5,sub6,sub7;

   dff #(3) subout0(in0, sub0, clk, reset);
   dff #(3) subout1(in1, sub1, clk, reset);
   dff #(3) subout2(in2, sub2, clk, reset);
   dff #(3) subout3(in3, sub3, clk, reset);
   dff #(3) subout4(in4, sub4, clk, reset);
   dff #(3) subout5(in5, sub5, clk, reset);
   dff #(3) subout6(in6, sub6, clk, reset);
   dff #(3) subout7(in7, sub7, clk, reset);

   function [2:0] subset_decode; input [2:0] a,b;
      begin
         subset_decode = 0;
         if (a<=b) subset_decode = a; else subset_decode = b;
      end
   endfunction
```

```
    function set_control; input [2:0] a,b;
       begin
          if (a<=b) set_control = 0; else set_control = 1;
       end
    endfunction
assign s0 = subset_decode (sub0,sub4);
assign s1 = subset_decode (sub1,sub5);
assign s2 = subset_decode (sub2,sub6);
assign s3 = subset_decode (sub3,sub7);
assign sout0 = set_control(sub0,sub4);
assign sout1 = set_control(sub1,sub5);
assign sout2 = set_control(sub2,sub6);
assign sout3 = set_control(sub3,sub7);
endmodule


/******************************************************/
/*    module compute_metric                          */
/******************************************************/
/* This module computes the sum of path memory and the distance for
each path entering a state of the trellis. For the four states,
there are two paths entering it; therefore eight sums are computed
in this module. The path metrics and output sums are 5 bits wide.
The output sum is bounded and should never be greater than 5 bits
for a valid input signal. The overflow from the sum is the error
output and indicates an invalid input signal.*/
module compute_metric
     (m_out0,m_out1,m_out2,m_out3,
     s0,s1,s2,s3,p0_0,p2_0,
     p0_1,p2_1,p1_2,p3_2,p1_3,p3_3,
     error);
   input [4:0] m_out0,m_out1,m_out2,m_out3;
   input [2:0] s0,s1,s2,s3;
   output [4:0] p0_0,p2_0,p0_1,p2_1,p1_2,p3_2,p1_3,p3_3;
   output error;

   assign
     p0_0 = m_out0 + s0,
     p2_0 = m_out2 + s2,
     p0_1 = m_out0 + s2,
     p2_1 = m_out2 + s0,
     p1_2 = m_out1 + s1,
     p3_2 = m_out3 + s3,
     p1_3 = m_out1 + s3,
     p3_3 = m_out3 + s1;

   function is_error; input x1,x2,x3,x4,x5,x6,x7,x8;
   begin
      if (x1||x2||x3||x4||x5||x6||x7||x8) is_error = 1;
```

```
        else is_error = 0;
      end
    endfunction

    assign error = is_error(p0_0[4],p2_0[4],p0_1[4],p2_1[4],
      p1_2[4],p3_2[4],p1_3[4],p3_3[4]);
endmodule

/*******************************************************/
/*    module compare_select                           */
/*******************************************************/
/* This module compares the summations from the compute_metric
module and selects the metric and path with the lowest value. The
output of this module is saved as the new path metric for each
state. The ACS output signals are used to control the path memory of
the decoder. */
module compare_select
      (p0_0,p2_0,p0_1,p2_1,p1_2,p3_2,p1_3,p3_3,
      out0,out1,out2,out3,
      ACS0,ACS1,ACS2,ACS3);
  input  [4:0] p0_0,p2_0,p0_1,p2_1,p1_2,p3_2,p1_3,p3_3;
  output [4:0] out0,out1,out2,out3;
  output ACS0,ACS1,ACS2,ACS3;

  function [4:0] find_min_metric; input [4:0] a,b;
    begin
      if (a <= b) find_min_metric = a; else find_min_metric = b;
    end
  endfunction

  function set_control; input [4:0] a,b;
    begin
      if (a <= b) set_control = 0; else set_control = 1;
    end
  endfunction
  assign out0 = find_min_metric(p0_0,p2_0);
  assign out1 = find_min_metric(p0_1,p2_1);
  assign out2 = find_min_metric(p1_2,p3_2);
  assign out3 = find_min_metric(p1_3,p3_3);
  assign ACS0 = set_control (p0_0,p2_0);
  assign ACS1 = set_control (p0_1,p2_1);
  assign ACS2 = set_control (p1_2,p3_2);
  assign ACS3 = set_control (p1_3,p3_3);
endmodule

/*******************************************************/
/*    module path                                     */
/*******************************************************/
/* This is the basic unit for the path memory of the Viterbi
```

decoder. It consists of four 3-bit D flip-flops in parallel. There
is a 2:1 mux at each D flip-flop input. The statement dff #(12)
instantiates a vector array of 12 flip-flops. */

```
module path(in,out,clk,reset,ACS0,ACS1,ACS2,ACS3);
input [11:0] in; output [11:0] out;
input clk,reset,ACS0,ACS1,ACS2,ACS3; wire [11:0] p_in;

dff #(12) path0(p_in,out,clk,reset);

   function [2:0] shift_path; input [2:0] a,b; input control;
     begin
        if (control == 0) shift_path = a; else shift_path = b;
     end
   endfunction

assign p_in[11:9] = shift_path(in[11:9],in[5:3],ACS0);
assign p_in[ 8:6] = shift_path(in[11:9],in[5:3],ACS1);
assign p_in[ 5:3] = shift_path(in[8: 6],in[2:0],ACS2);
assign p_in[ 2:0] = shift_path(in[8: 6],in[2:0],ACS3);
endmodule


/****************************************************/
/*    module path_memory                            */
/****************************************************/
/* This module consists of an array of memory elements (D
flip-flops) that store and shift the path memory as new signals are
added to the four paths (or four most likely sequences of signals).
This module instantiates 11 instances of the path module. */
module path_memory
     (p0,p1,p2,p3,
     path0,clk,reset,
     ACS0,ACS1,ACS2,ACS3);
output [2:0] p0,p1,p2,p3; input [11:0] path0;
input clk,reset,ACS0,ACS1,ACS2,ACS3;
wire [11:0]out1,out2,out3,out4,out5,out6,out7,out8,out9,out10,out11;
     path x1 (path0,out1 ,clk,reset,ACS0,ACS1,ACS2,ACS3),
          x2 (out1, out2 ,clk,reset,ACS0,ACS1,ACS2,ACS3),
          x3 (out2, out3 ,clk,reset,ACS0,ACS1,ACS2,ACS3),
          x4 (out3, out4 ,clk,reset,ACS0,ACS1,ACS2,ACS3),
          x5 (out4, out5 ,clk,reset,ACS0,ACS1,ACS2,ACS3),
          x6 (out5, out6 ,clk,reset,ACS0,ACS1,ACS2,ACS3),
          x7 (out6, out7 ,clk,reset,ACS0,ACS1,ACS2,ACS3),
          x8 (out7, out8 ,clk,reset,ACS0,ACS1,ACS2,ACS3),
          x9 (out8, out9 ,clk,reset,ACS0,ACS1,ACS2,ACS3),
          x10(out9, out10,clk,reset,ACS0,ACS1,ACS2,ACS3),
          x11(out10,out11,clk,reset,ACS0,ACS1,ACS2,ACS3);
assign p0 = out11[11:9];
assign p1 = out11[ 8:6];
assign p2 = out11[ 5:3];
```

```
      assign p3 = out11[ 2:0];
      endmodule


/**********************************************************/
/*    module pathin                                       */
/**********************************************************/
/* This module determines the input signal to the path for each of
the four paths. Control signals from the subset decoder and compare
select modules are used to store the correct signal. The statement
dff #(12) instantiates a vector array of 12 flip-flops. */
module pathin
      (sout0,sout1,sout2,sout3,
      ACS0,ACS1,ACS2,ACS3,
      path0,clk,reset);
   input sout0,sout1,sout2,sout3,ACS0,ACS1,ACS2,ACS3;
   input clk,reset; output [11:0] path0;
   wire [2:0] sig0,sig1,sig2,sig3; wire [11:0] path_in;

   dff #(12) firstpath(path_in,path0,clk,reset);

   function [2:0] subset0; input sout0;
      begin
         if(sout0 == 0) subset0 = 0; else subset0 = 4;
      end
   endfunction

   function [2:0] subset1; input sout1;
      begin
         if(sout1 == 0) subset1 = 1; else subset1 = 5;
      end
   endfunction

   function [2:0] subset2; input sout2;
      begin
         if(sout2 == 0) subset2 = 2; else subset2 = 6;
      end
   endfunction

   function [2:0] subset3; input sout3;
      begin
         if(sout3 == 0) subset3 = 3; else subset3 = 7;
      end
   endfunction

   function [2:0] find_path; input [2:0] a,b; input control;
      begin
         if(control==0) find_path = a; else find_path = b;
      end
   endfunction

assign sig0 = subset0(sout0);
```

```
assign sig1 = subset1(sout1);
assign sig2 = subset2(sout2);
assign sig3 = subset3(sout3);
assign path_in[11:9] = find_path(sig0,sig2,ACS0);
assign path_in[ 8:6] = find_path(sig2,sig0,ACS1);
assign path_in[ 5:3] = find_path(sig1,sig3,ACS2);
assign path_in[ 2:0] = find_path(sig3,sig1,ACS3);
endmodule

/****************************************************/
/*    module metric                                 */
/****************************************************/
/* The registers created in this module (using D flip-flops) store
the four path metrics. Each register is 5 bits wide. The statement
dff #(5) instantiates a vector array of 5 flip-flops. */
module metric
     (m_in0,m_in1,m_in2,m_in3,
     m_out0,m_out1,m_out2,m_out3,
     clk,reset);
input [4:0] m_in0,m_in1,m_in2,m_in3;
output [4:0] m_out0,m_out1,m_out2,m_out3;
input clk,reset;
   dff #(5) metric3(m_in3, m_out3, clk, reset);
   dff #(5) metric2(m_in2, m_out2, clk, reset);
   dff #(5) metric1(m_in1, m_out1, clk, reset);
   dff #(5) metric0(m_in0, m_out0, clk, reset);
endmodule

/****************************************************/
/*    module output_decision                        */
/****************************************************/
/* This module decides the output signal based on the path that
corresponds to the smallest metric. The control signal comes from
the reduce module. */

module output_decision(p0,p1,p2,p3,control,out);
   input [2:0] p0,p1,p2,p3; input [1:0] control; output [2:0] out;
   function [2:0] decide;
   input [2:0] p0,p1,p2,p3; input [1:0] control;
   begin
     if(control == 0) decide = p0;
     else if(control == 1) decide = p1;
     else if(control == 2) decide = p2;
     else decide = p3;
     end
   endfunction
assign out = decide(p0,p1,p2,p3,control);
endmodule
```

```
/********************************************************/
/*    module reduce                                     */
/********************************************************/
/* This module reduces the metrics after the addition and compare
operations. This algorithm selects the smallest metric and subtracts
it from all the other metrics. */
module reduce
    (in0,in1,in2,in3,
    m_in0,m_in1,m_in2,m_in3,
    control);
  input [4:0] in0,in1,in2,in3;
  output [4:0] m_in0,m_in1,m_in2,m_in3;
  output [1:0] control; wire [4:0] smallest;

  function [4:0] find_smallest;
    input [4:0] in0,in1,in2,in3; reg [4:0] a,b;
      begin
        if(in0 <= in1) a = in0; else a = in1;
        if(in2 <= in3) b = in2; else b = in3;
        if(a <= b) find_smallest = a;
        else find_smallest = b;
      end
  endfunction

  function [1:0] smallest_no;
  input [4:0] in0,in1,in2,in3,smallest;
    begin
      if(smallest == in0) smallest_no = 0;
      else if (smallest == in1) smallest_no = 1;
      else if (smallest == in2) smallest_no = 2;
      else smallest_no = 3;
    end
  endfunction

assign smallest = find_smallest(in0,in1,in2,in3);
assign m_in0 = in0 - smallest;
assign m_in1 = in1 - smallest;
assign m_in2 = in2 - smallest;
assign m_in3 = in3 - smallest;
assign control = smallest_no(in0,in1,in2,in3,smallest);
endmodule
```

11.13 Other Verilog Features

This section covers some of the more advanced Verilog features. **System tasks** and functions are defined as part of the IEEE Verilog standard [Verilog LRM14].

11.13.1   Display Tasks

The following code illustrates the **display system tasks** [Verilog LRM 14.1]:

```
module test_display; // display system tasks:
initial begin $display ("string, variables, or expression");
/* format specifications work like printf in C:
          %d=decimal %b=binary %s=string %h=hex %o=octal
          %c=character %m=hierarchical name %v=strength %t=time format
          %e=scientific %f=decimal %g=shortest
examples: %d uses default width %0d uses minimum width
          %7.3g uses 7 spaces with 3 digits after decimal point */
// $displayb, $displayh, $displayo print in b, h, o formats
// $write, $strobe, $monitor also have b, h, o versions

$write("write"); // as $display, but without newline at end of line

$strobe("strobe"); // as $display, values at end of simulation cycle

$monitor(v); // disp. @change of v (except v= $time,$stime,$realtime)
$monitoron; $monitoroff; // toggle monitor mode on/off

end endmodule
```

11.13.2   File I/O Tasks

The following example illustrates the **file I/O system tasks** [Verilog LRM 14.2]:

```
module file_1; integer f1, ch; initial begin f1 = $fopen("f1.out");
if(f1==0) $stop(2); if(f1==2)$display("f1 open");
ch = f1|1; $fdisplay(ch,"Hello"); $fclose(f1); end endmodule

> vlog file_1.v
> vsim -c file_1
# Loading work.file_1
VSIM 1> run 10
# f1 open
# Hello
VSIM 2> q
> more f1.out
Hello
>
```

The $fopen system task returns a 32-bit unsigned integer called a **multichannel descriptor** (f1 in this example) unique to each file. The multichannel descriptor contains 32 flags, one for each of 32 possible channels or files (subject to limitations of the operating system). Channel 0 is the standard output (normally the screen), which is always open. The first call to $fopen opens channel 1 and sets bit 1 of the multichannel descriptor. Subsequent calls set higher bits. The file I/O system tasks: $fdisplay, $fwrite, $fmonitor, and $fstrobe; correspond to their display counterparts. The first parameter for the file system tasks is a multichannel descriptor that may have

.multiple bits set. Thus, the preceding example writes the string `"Hello"` to the screen and to `file1.out`. The task `$fclose` closes a file and allows the channel to be reused.

The file I/O tasks `$readmemb` and `$readmemh` read a text file into a memory. The file may contain only spaces, new lines, tabs, form feeds, comments, addresses, and binary (for `$readmemb`) or hex (for `$readmemh`) numbers, as in the following example:

```
mem.dat
@2 1010_1111 @4 0101_1111 1010_1111 // @address in hex
x1x1_zzzz 1111_0000 /* x or z is OK */
```

```
module load; reg [7:0] mem[0:7]; integer i; initial begin
$readmemb("mem.dat", mem, 1, 6); // start_address=1, end_address=6
for (i= 0; i<8; i=i+1) $display("mem[%0d] %b", i, mem[i]);
end endmodule
```

```
> vsim -c load
# Loading work.load
VSIM 1> run 10
# ** Warning: $readmem (memory mem) file mem.dat line 2:
#    More patterns than index range (hex 1:6)
#    Time: 0 ns  Iteration: 0  Instance:/
# mem[0] xxxxxxxx
# mem[1] xxxxxxxx
# mem[2] 10101111
# mem[3] xxxxxxxx
# mem[4] 01011111
# mem[5] 10101111
# mem[6] x1x1zzzz
# mem[7] xxxxxxxx
VSIM 2> q
>
```

11.13.3 Timescale, Simulation, and Timing-Check Tasks

There are two **timescale tasks**, `$printtimescale` and `$timeformat` [Verilog LRM 14.3]. The `$timeformat` specifies the `%t` format specification for the display and file I/O system tasks as well as the time unit for delays entered interactively and from files. Here are examples of the timescale tasks:

```
// timescale tasks:
module a; initial $printtimescale(b.c1); endmodule
module b; c c1 (); endmodule
`timescale 10 ns / 1 fs
module c_dat; endmodule

`timescale 1 ms / 1 ns
module Ttime; initial $timeformat(-9, 5, " ns", 10); endmodule
/* $timeformat [ ( n, p, suffix , min_field_width ) ] ;
```

```
units = 1 second ** (-n), n = 0->15, e.g. for n = 9, units = ns
p = digits after decimal point for %t e.g. p = 5 gives 0.00000
suffix for %t (despite timescale directive)
min_field_width is number of character positions for %t */
```

The **simulation control tasks** are $stop and $finish [Verilog LRM 14.4]:

```
module test_simulation_control; // simulation control system tasks:
initial begin $stop; // enter interactive mode (default parameter 1)
$finish(2); // graceful exit with optional parameter as follows:
// 0 = nothing 1 = time and location 2 = time, location, and statistics
end endmodule
```

The **timing-check tasks** [Verilog LRM 14.5] are used in specify blocks. The following code and comments illustrate the definitions and use of timing-check system tasks. The arguments to the tasks are defined and explained in Table 11.11.

TABLE 11.11   Timing-check system task parameters.

Timing task argument	Description of argument	Type of argument
reference_event	to establish reference time	module input or inout (scalar or vector net)
data_event	signal to check against reference_event	module input or inout (scalar or vector net)
limit	time limit to detect timing violation on data_event	constant expression or specparam
threshold	largest pulse width ignored by timing check $width	constant expression or specparam
notifier	flags a timing violation (before -> after): x->0, 0->1, 1->0, z->z	register

```
module timing_checks (data, clock, clock_1,clock_2);              //1
input data,clock,clock_1,clock_2; reg tSU,tH,tHIGH,tP,tSK,tR;     //2
specify // timing check system tasks:                             //3
/* $setup (data_event, reference_event, limit [, notifier]);      //4
violation = (T_reference_event)-(T_data_event) < limit */         //5
$setup(data, posedge clock, tSU);                                 //6
/* $hold (reference_event, data_event, limit [, notifier]);       //7
violation =                                                       //8
  (time_of_data_event)-(time_of_reference_event) < limit */       //9
$hold(posedge clock, data, tH);                                   //10
/* $setuphold (reference_event, data_event, setup_limit,          //11
    hold_limit [, notifier]);                                     //12
parameter_restriction = setup_limit + hold_limit > 0 */           //13
```

```
$setuphold(posedge clock, data, tSU, tH);                              //14
/* $width (reference_event, limit, threshold [, notifier]);            //15
violation =                                                            //16
   threshold < (T_data_event) - (T_reference_event) < limit            //17
reference_event = edge                                                 //18
data_event = opposite_edge_of_reference_event */                       //19
$width(posedge clock, tHIGH);                                          //20
/* $period (reference_event, limit [, notifier]);                      //21
violation = (T_data_event) - (T_reference_event) < limit               //22
reference_event = edge                                                 //23
data_event = same_edge_of_reference event */                           //24
$period(posedge clock, tP);                                            //25
/* $skew (reference_event, data_event, limit [, notifier]);            //26
violation = (T_data_event) - (T_reference_event) > limit */            //27
$skew(posedge clock_1, posedge clock_2, tSK);                          //28
/* $recovery (reference_event, data_event, limit, [, notifier]);       //29
violation = (T_data_event) - (T_reference_event) < limit */            //30
$recovery(posedge clock, posedge clock_2, tR);                         //31
/* $nochange (reference_event, data_event, start_edge_offset,          //32
   end_edge_offset [, notifier]);                                      //33
reference_event = posedge | negedge                                    //34
violation = change while reference high (posedge)/low (negedge)        //35
+ve start_edge_offset moves start of window later                      //36
+ve end_edge_offset moves end of window later */                       //37
$nochange (posedge clock, data, 0, 0);                                 //38
endspecify endmodule                                                   //39
```

You can use **edge specifiers** as parameters for the timing-check events (except for the reference event in $nochange):

```
edge_control_specifier ::= edge [edge_descriptor {, edge_descriptor}]
edge_descriptor ::= 01 | 0x | 10 | 1x | x0 | x1
```

For example, 'edge [01, 0x, x1] clock' is equivalent to 'posedge clock'. Edge transitions with 'z' are treated the same as transitions with 'x'.

Here is a D flip-flop model that uses timing checks and a **notifier register**. The register, notifier, is changed when a timing-check task detects a violation and the last entry in the table then sets the flip-flop output to unknown.

```
primitive dff_udp(q, clock, data, notifier);
output q; reg q; input clock, data, notifier;
table //    clock data  notifier:state:  q
            r    0     ?       : ? :  0 ;
            r    1     ?       : ? :  1 ;
            n    ?     ?       : ? :  - ;
            ?    *     ?       : ? :  - ;
            ?    ?     *       : ? :  x ; endtable // notifier
```

```
endprimitive

`timescale 100 fs / 1 fs
module dff(q, clock, data); output q; input clock, data; reg notifier;
dff_udp(q1, clock, data, notifier); buf(q, q1);
specify
    specparam tSU = 5, tH = 1, tPW = 20, tPLH = 4:5:6, tPHL = 4:5:6;
        (clock *> q) = (tPLH, tPHL);
    $setup(data, posedge clock, tSU, notifier); // setup: data to clock
    $hold(posedge clock, data, tH, notifier); // hold: clock to data
    $period(posedge clock, tPW, notifier); // clock: period
endspecify
endmodule
```

11.13.4 PLA Tasks

The **PLA modeling tasks** model two-level logic [Verilog LRM 14.6]. As an example, the following eqntott logic equations can be implemented using a PLA:

```
b1 = a1 & a2; b2 = a3 & a4 & a5 ; b3 = a5 & a6 & a7;
```

The following module loads a PLA model for the equations above (in AND logic) using the **array format** (the array format allows only '1' or '0' in the PLA memory, or **personality array**). The file array.dat is similar to the espresso input plane format.

```
array.dat
1100000
0011100
0000111

module pla_1 (a1,a2,a3,a4,a5,a6,a7,b1,b2,b3);
input a1, a2, a3, a4, a5, a6, a7 ; output b1, b2, b3;
reg [1:7] mem[1:3]; reg b1, b2, b3;
initial begin
    $readmemb("array.dat", mem);
    #1; b1=1; b2=1; b3=1;
    $async$and$array(mem,{a1,a2,a3,a4,a5,a6,a7},{b1,b2,b3});
end
initial $monitor("%4g",$time,,b1,,b2,,b3);
endmodule
```

The next example illustrates the use of the **plane format**, which allows '1', '0', as well as '?' or 'z' (either may be used for don't care) in the personality array.

```
b1 = a1 & !a2; b2 = a3; b3 = !a1 & !a3; b4 = 1;

module pla_2; reg [1:3] a, mem[1:4]; reg [1:4] b;
initial begin
    $async$and$plane(mem,{a[1],a[2],a[3]},{b[1],b[2],b[3],b[4]});
    mem[1] = 3'b10?; mem[2] = 3'b??1; mem[3] = 3'b0?0; mem[4] = 3'b???;
```

```
        #10 a = 3'b111; #10 $displayb(a, " -> ", b);
        #10 a = 3'b000; #10 $displayb(a, " -> ", b);
        #10 a = 3'bxxx; #10 $displayb(a, " -> ", b);
        #10 a = 3'b101; #10 $displayb(a, " -> ", b);
end endmodule
111 -> 0101
000 -> 0011
xxx -> xxx1
101 -> 1101
```

11.13.5 Stochastic Analysis Tasks

The **stochastic analysis tasks** model queues [Verilog LRM 14.7]. Each of the tasks return a status as shown in Table 11.12.

TABLE 11.12 Status values for the stochastic analysis tasks.

Status value	Meaning
0	OK
1	queue full, cannot add
2	undefined q_id
3	queue empty, cannot remove
4	unsupported q_type, cannot create queue
5	max_length <= 0, cannot create queue
6	duplicate q_id, cannot create queue
7	not enough memory, cannot create queue

The following module illustrates the interface and parameters for these tasks:

```
module stochastic; initial begin // stochastic analysis system tasks:

/* $q_initialize (q_id, q_type, max_length, status) ;
q_id is an integer that uniquely identifies the queue
q_type 1=FIFO 2=LIFO
max_length is an integer defining the maximum number of entries */
$q_initialize (q_id, q_type, max_length, status) ;

/* $q_add (q_id, job_id, inform_id, status) ;
job_id = integer input
inform_id = user-defined integer input for queue entry */
$q_add (q_id, job_id, inform_id, status) ;

/* $q_remove (q_id, job_id, inform_id, status) ; */
$q_remove (q_id, job_id, inform_id, status) ;
```

```
/* $q_full (q_id, status) ;
status = 0 = queue is not full, status = 1 = queue full */
$q_full (q_id, status) ;

/* $q_exam (q_id, q_stat_code, q_stat_value, status) ;
q_stat_code is input request as follows:
1=current queue length 2=mean inter-arrival time 3=max. queue length
4=shortest wait time ever
5=longest wait time for jobs still in queue 6=ave. wait time in queue
q_stat_value is output containing requested value */
$q_exam (q_id, q_stat_code, q_stat_value, status) ;

end endmodule
```

11.13.6 Simulation Time Functions

The **simulation time functions** return the time as follows [Verilog LRM 14.8]:

```
module test_time; initial begin // simulation time system functions:
$time ;
// returns 64-bit integer scaled to timescale unit of invoking module

$stime ;
// returns 32-bit integer scaled to timescale unit of invoking module

$realtime ;
// returns real scaled to timescale unit of invoking module

end endmodule
```

11.13.7 Conversion Functions

The **conversion functions for reals** handle real numbers [Verilog LRM 14.9]:

```
module test_convert; // conversion functions for reals:
integer i; real r; reg [63:0] bits;
initial begin #1 r=256;#1 i = $rtoi(r);
#1; r = $itor(2 * i) ; #1 bits = $realtobits(2.0 * r) ;
#1; r = $bitstoreal(bits) ; end
initial $monitor("%3f",$time,,i,,r,,bits); /*
$rtoi converts reals to integers w/truncation e.g. 123.45 -> 123
$itor converts integers to reals e.g. 123 -> 123.0
$realtobits converts reals to 64-bit vector
$bitstoreal converts bit pattern to real
Real numbers in these functions conform to IEEE Std 754. Conversion
rounds to the nearest valid number. */
endmodule
```

```
# 0.000000          x 0              x
# 1.000000          x 256                x
# 2.000000        256 256                x
# 3.000000        256 512                x
```

```
# 4.000000        256  512   4652218415073722368
# 5.000000        256  1024  4652218415073722368
```

Here is an example using the conversion functions in port connections:

```
module test_real;wire [63:0]a; driver d (a); receiver r (a);
initial $monitor("%3g",$time,,a,,d.r1,,r.r2); endmodule

module driver (real_net);
output real_net; real r1; wire [64:1] real_net = $realtobits(r1);
initial #1 r1 = 123.456; endmodule

module receiver (real_net);
input real_net; wire [64:1] real_net; real r2;
initial assign r2 = $bitstoreal(real_net);
endmodule

# 0                         0 0 0
# 1   4638387860618067575 123.456 123.456
```

11.13.8 Probability Distribution Functions

The probability distribution functions are as follows [Verilog LRM 14.10]:

```
module probability; // probability distribution functions:          //1
/* $random [(seed)] returns random 32-bit signed integer            //2
seed = register, integer, or time */                                //3
reg [23:0] r1,r2; integer r3,r4,r5,r6,r7,r8,r9;                      //4
integer seed, start, \end , mean, standard_deviation;               //5
integer degree_of_freedom, k_stage;                                 //6
initial begin seed=1; start=0; \end =6; mean=5;                     //7
standard_deviation=2; degree_of_freedom=2; k_stage=1; #1;           //8
r1 = $random % 60; // random -59 to 59                              //9
r2 = {$random} % 60; // positive value 0-59                         //10
r3=$dist_uniform (seed, start, \end ) ;                             //11
r4=$dist_normal (seed, mean, standard_deviation) ;                  //12
r5=$dist_exponential (seed, mean) ;                                 //13
r6=$dist_poisson (seed, mean) ;                                     //14
r7=$dist_chi_square (seed, degree_of_freedom) ;                     //15
r8=$dist_t (seed, degree_of_freedom) ;                              //16
r9=$dist_erlang (seed, k_stage, mean) ; end                         //17
initial #2 $display ("%3f",$time,,r1,,r2,,r3,,r4,,r5);              //18
initial begin #3; $display ("%3f",$time,,r6,,r7,,r8,,r9); end       //19
/* All parameters are integer values.                               //20
Each function returns a pseudo-random number                        //21
e.g. $dist_uniform returns uniformly distributed random numbers     //22
mean, degree_of_freedom, k_stage                                    //23
(exponential, poisson, chi-square, t, erlang) > 0.                  //24
seed = inout integer initialized by user, updated by function       //25
```

```
start, end ($dist_uniform) = integer bounding return values */      //26
endmodule                                                           //27
2.000000        8       57        0         4          9
3.000000              7        3         0        2
```

11.13.9 Programming Language Interface

The C language **Programming Language Interface** (**PLI**) allows you to access the internal Verilog data structure [Verilog LRM17–23, A–E]. For example, you can use the PLI to implement the following extensions to a Verilog simulator:

- C language delay calculator for a cell library
- C language interface to a Verilog-based or other logic or fault simulator
- Graphical waveform display and debugging
- C language simulation models
- Hardware interfaces

There are three generations of PLI routines (see Appendix B for an example):

- Task/function (TF) routines (or utility routines), the first generation of the PLI, start with `'tf_'`.
- Access (ACC) routines, the second generation of the PLI, start with the characters `'acc_'` and access delay and logic values. There is some overlap between the ACC routines and TF routines.
- Verilog Procedural Interface (VPI) routines, the third generation of the PLI, start with the characters `'vpi_'` and are a superset of the TF and ACC routines.

11.14 Summary

Table 11.13 lists the key features of Verilog HDL. The most important concepts covered in this chapter are:

- Concurrent processes and sequential execution
- Difference between a `reg` and a `wire`, and between a scalar and a vector
- Arithmetic operations on `reg` and `wire`
- Data slip
- Delays and events

TABLE 11.13 Verilog on one page.

Verilog feature	Example

Comments	`a = 0; // comment ends with newline`
	`/* This is a multiline or block`
	`comment */`

Constants: string and numeric	`parameter BW = 32 // local, use BW`
	`` `define G_BUS 32 // global, use `G_BUS ``
	`4'b2  1'bx`

Names (case-sensitive, start with letter or '_')	`_12name  A_name  $BAD  NotSame  notsame`
Two basic types of logic signals: wire and reg	`wire myWire; reg myReg;`
Use a continuous assignment statement with wire	`assign myWire = 1;`
Use a procedural assignment statement with reg	`always myReg = myWire;`
Buses and vectors use square brackets	`reg [31:0] DBus; DBus[12] = 1'bx;`
We can perform arithmetic on bit vectors	`reg [31:0] DBus; DBus = DBus + 2;`
Arithmetic is performed modulo 2^n	`reg [2:0] R; R = 7 + 1; // now R = 0`
Operators: as in C (but not ++ or --)	
Fixed logic-value system	`1, 0, x` (unknown), `z` (high-impedance)

Basic unit of code is the module	`module bake (chips, dough, cookies);`
	`input chips, dough; output cookies;`
	`assign cookies = chips & dough;`
	`endmodule`

Ports	input or input/output ports are `wire` output ports are `wire` or `reg`

Procedures model things that happen at the same time and may be sensitive to an edge, **posedge**, **negedge**, or to a level.	`always @rain sing; always @rain dance;`
	`always @(posedge clock) D = Q; // flop`
	`always @(a or b) c = a & b; // and gate`

Sequential blocks model repeating things:
always: executes forever
initial: executes once only at start of simulation
	`initial born;`
	`always @alarm_clock begin : a_day`
	`metro=commute; boulot=work; dodo=sleep;`
	`end`

Functions and tasks	`function ... endfunction` `task ... endtask`
Output	`$display("a=%f",a);$dumpvars;$monitor(a)`
Control simulation	`$stop; $finish // sudden or gentle halt`
Compiler directives	`` `timescale 1ns/1ps // units/resolution ``

Delay	`#1 a = b;  // delay then sample b`
	`a = #1 b;  // sample b then delay`

11.15 Problems

*=Difficult, **=Very difficult, ***=Extremely difficult

11.1 (Counter, 30 min.) Download the VeriWell simulator from `http://www.wellspring.com` and simulate the counter from Section 11.1 (exclude the comments to save typing). Include the complete input and output listings in your report.

11.2 (Simulator, 30 min.) Build a "cheat sheet" for your simulator, listing the commands for running the simulator and using it in interactive mode.

11.3 (Verilog examples, 10 min.) The Cadence Verilog-XL simulator comes with a directory `examples`. Make a list of the examples from the `README` files in the various directories.

11.4 (Gotchas, 60 min.) Build a "most common Verilog mistakes" file. Start with:

- Extra or missing semicolon ';'
- Forgetting to declare a `reg`
- Using a `reg` instead of a `wire` for an `input` or `inout` port
- Bad declarations: `reg bus[0:31]` instead of `reg [31:0]bus`
- Mixing vector declarations: `wire [31:0]BusA, [15:0]BusB`
- The case-sensitivity of Verilog
- No delay in an `always` statement (simulator loops forever)
- Mixing up ` (accent grave) for `` `define `` and ' (tick or apostrophe) for `1'b1` with ´ (accent acute) or ' (open single quote) or ' (close single quote)
- Mixing " (double quote) with " (open quotes) or " (close quotes)

11.5 (Sensitivity, 10 min.) Explore and explain what happens if you write this:

```
always @(a or b or c) e = (a|b)&(c|d);
```

11.6 (Verilog `if` statement, 10 min.) Build test code to simulate the following Verilog fragment. Explain what is wrong and fix the problem.

```
if (i > 0)
  if (i < 2) $display ("i is 1");
else $display ("i is less than 0");
```

11.7 (Effect of delay, 30 min.). Write code to test the four different code fragments shown in Table 11.14 and print the value of `'a'` at time = 0 and time = 1 for each case. Explain the differences in your simulation results.

11.8 (Verilog events, 10 min.). Simulate the following and explain the results:

```
event event_1, event_2;
always @ event_1 -> event_2;
initial @event_2 $stop;
initial -> event_1;
```

TABLE 11.14 Code fragments for Problem 11.7.

	(a)	(b)	(c)	(d)

Code fragment	`reg a;`
	`initial`
	`begin`
	`a = 0;`
	`a = a + 1;`
	`end`
		`reg a;`
		`initial`
		`begin`
		`#0 a = 0;`
		`#0 a = a + 1;`
		`end`
			`reg a;`
			`initial`
			`begin`
			`a <= 0;`
			`a <= a + 1;`
			`end`
				`reg a;`
				`initial`
				`begin`
				`#1 a = 0;`
				`#1 a = a + 1;`
				`end`

11.9 (Blocking and nonblocking assignment statements, 30 min.). Write code to test the different code fragments shown in Table 11.15 and print the value of 'outp' at time = 0 and time = 10 for each case. Explain the difference in simulation results.

TABLE 11.15 Code fragments for Problem 11.9.

	(a)	(b)	(c)	(d)

Code fragment	`reg outp;`
	`always`
	`begin`
	`#10 outp = 0;`
	`#10 outp = 1;`
	`end`
		`reg outp;`
		`always`
		`begin`
		`outp <= #10 1;`
		`outp <= #10 0;`
		`end`
			`reg outp;`
			`always`
			`begin`
			`#10 outp = 0;`
			`#10 outp <= 1;`
			`end`
				`reg outp;`
				`always`
				`begin`
				`#10 outp <= 0;`
				`#10 outp = 1;`
				`end`

11.10 (Verilog UDPs, 20 min.). Use this primitive to build a half adder:

```
primitive Adder(Sum, InA, InB); output Sum; input Ina, InB;
table 00 : 0; 01 : 1; 10 : 1; 11 : 0; endtable
endprimitive
```

Apply unknowns to the inputs. What is the output?

11.11 (Verilog UDPs, 30 min.). Use the following primitive model for a D latch:

```
primitive DLatch(Q, Clock, Data); output Q; reg Q; input Clock, Data;
table 1 0 : ? : 0; 1 1 : ? : 1; 0 1 : ? : -; endtable
endprimitive
```

Check to see what happens when you apply unknown inputs (including clock transitions to unknown). What happens if you apply high-impedance values to the inputs (again including transitions)?

11.12 (Propagation of unknowns in primitives, 45 min.) Use the following primitive model for a D flip-flop:

```
primitive DFF(Q, Clock, Data); output Q; reg Q; input Clock, Data;
```

```
table
r    0 : ? : 0 ;
r    1 : ? : 1 ;
(0x) 0 : 0 : 0 ;
(0x) 1 : 1 : 1 ;
(?0) ? : ? : - ;
? (??) : ? : - ;
endtable
endprimitive
```

Check to see what happens when you apply unknown inputs (including a clock transition to an unknown value). What happens if you apply high-impedance values to the inputs (again including transitions)?

11.13 (D flip-flop UDP, 60 min.) Table 11.16 shows a UDP for a D flip-flop with QN output and asynchronous reset and set.

TABLE 11.16   D flip-flop UDP for Problem 11.13.

```
primitive DFlipFlop2(QN, Data, Clock, Res, Set);
output QN; reg QN; input Data, Clock, Res, Set;
table
//      Data    Clock   Res     Set     :state  :next state
        1       (01)    0       0       :?      :0;     // line 1
        1       (01)    0       x       :?      :0;
        ?       ?       0       x       :0      :0;
        0       (01)    0       0       :?      :1;
        0       (01)    x       0       :?      :1;
        ?       ?       x       0       :1      :1;
        1       (x1)    0       0       :0      :0;
        0       (x1)    0       0       :1      :1;
        1       (0x)    0       0       :0      :0;
        0       (0x)    0       0       :1      :1;
        ?       ?       1       ?       :?      :1;
        ?       ?       0       1       :?      :0;
        ?       n       0       0       :?      :-;
        *       ?       ?       ?       :?      :-;
        ?       ?       (?0)    ?       :?      :-;
        ?       ?       ?       (?0)    :?      :-;
        ?       ?       ?       ?       :?      :x;     // line 17
endtable
endprimitive
```

---

a. Explain the purpose of each line in the truth table.

b. Write a module to test each line of the UDP.

c. Can you find any errors, omissions, or other problems in this UDP?

11.14 (JK flip-flop, 30 min.) Test the following model for a JK flip-flop:

```
module JKFF (Q, J, K, Clk, Rst);
parameter width = 1, reset_value = 0;
input [width-1:0] J, K; output [width-1:0] Q; reg [width-1:0] Q;
input Clk, Rst; initial Q = {width{1'bx}};
always @ (posedge Clk or negedge Rst )
if (Rst==0 ) Q <= #1 reset_value;
else Q <= #1 (J & ~K) | (J & K & ~Q) | (~J & ~K & Q);
endmodule
```

11.15 (Overriding Verilog parameters, 20 min.) The following module has a parameter specification that allows you to change the number of AND gates that it models (the cardinality or width):

```
module Vector_AND(Z, A, B);
   parameter card = 2; input [card-1:0] A,B; output [card-1:0] Z;
   wire [card-1:0] Z = A & B;
endmodule
```

The next module changes the parameter value by specifying an overriding value in the module instantiation:

```
module Four_AND_Gates(OutBus, InBusA, InBusB);
   input [3:0] InBusA, InBusB; output [3:0] OutBus;
   Vector_AND #(4) My_AND(OutBus, InBusA, InBusB);
endmodule
```

These next two modules change the parameter value by using a defparam statement, which overrides the declared parameter value:

```
module X_AND_Gates(OutBus, InBusA, InBusB);
   parameter X = 2;input [X-1:0] InBusA, InBusB;output [X-1:0] OutBus;
   Vector_AND #(X) My_AND(OutBus, InBusA, InBusB);
endmodule

module size; defparam X_AND_Gates.X = 4; endmodule
```

    a. Check that the two alternative methods of specifying parameters are equivalent by instantiating the modules Four_AND_Gates and X_AND_Gates in another module and simulating.

    b. List and comment on the advantages and disadvantages of both methods.

11.16 (Default Verilog delays, 10 min.). Demonstrate, using simulation, that the following NAND gates have the delays you expect:

```
nand (strong0, strong1) #1
   Nand_1(n001, n004, n005),
   Nand_2(n003, n001, n005, n002);
nand (n006, n005, n002);
```

11.17 (Arrays of modules, 30 min.) Newer versions of Verilog allow the instantiating of **arrays of modules** (in this book we usually call this a vector since we are only allowed one row). You specify the number in the array by using a **range** after the instance name as follows:

```
nand #2 nand_array[0:7](zn, a, b);
```

Create and test a model for an 8-bit register using an array of flip-flops.

11.18 (Assigning Verilog real to integer data types, 10 min.). What is the value of ImInteger in the following code?

```
real ImReal; integer ImInteger;
initial begin ImReal = -1.5; ImInteger = ImReal; end
```

11.19 (BNF syntax, 10 min.) Use the BNF syntax definitions in Appendix B to answer the following questions. In each case explain how you arrive at the answer:

a. What is the highest-level construct?

b. What is the lowest-level construct?

c. Can you nest begin and end statements?

d. Where is a legal place for a case statement?

e. Is the following code legal: **reg** [31:0] rega, [32:1] regb;

f. Where is it legal to include sequential statements?

11.20 (Old syntax definitions, 10 min.) Prior to the IEEE LRM, Verilog BNF was expressed using a different notation. For example, an event expression was defined as follows:

```
<event_expression> ::= <expression>
  or <<posedge or negedge> <SCALAR_EVENT_EXPRESSION>>
  or <<event_expression> or <event_expression>>
```

Notice that we are using 'or' as part of the BNF to mean "alternatively" and also '**or**' as a Verilog keyword. The keyword '**or**' is in bold—the difference is fairly obvious. Here is an alternative definition for an event expression:

```
<event_expression> ::= <expression>
||= posedge <SCALAR_EVENT_EXPRESSION>
||= negedge <SCALAR_EVENT_EXPRESSION>
||= <event_expression> <or <event_expression>>*
```

Are these definitions equivalent (given, of course, that we replaced ||= with or in the simplified syntax)? Explain carefully how you would attempt to prove that they are the same.

11.21 (Operators, 20 min.) Explain Table 11.17 (see next page).

11.22 (Unary reduction, 10 min.) Complete Table 11.18 (see next page).

11.23 (Coerced ports, 20 min.) Perform some experiments to test the behavior of your Verilog simulator in the following situation: "NOTE—A port that is declared as input (output) but used as an output (input) or inout may be coerced to inout. If not coerced to inout, a warning must be issued" [Verilog LRM 12.3.6].

TABLE 11.17 Unary operators (Problem 11.21).

	(a)	(b)	(c)	(d)

Code	```module unary;```
	```reg [4:0] u;```
	```initial u=!'b011z;```
	```initial```
	```$display("%b",u);```
	```endmodule```
		```module unary;```
		```wire u;```
		```assign u=!'b011z;```
		```initial```
		```$display("%b",u);```
		```endmodule```
			```module unary;```
			```wire u;```
			```assign u=!'b011z;```
			```initial```
			```#1 $display("%b",u);```
			```endmodule```
				```module unary;```
				```wire u;```
				```assign u=&'b1;```
				```initial```
				```#1 $display("%b",u);```
				```endmodule```

Output	0000x	z	x	0

TABLE 11.18 Unary reduction (Problem 11.22).

Operand	&	~&	|	~|	^	~^
4'b0000						
4'b1111						
4'b01x0						
4'bz000						

11.24 (*Difficult delay code, 20 min.) Perform some experiments to explain what this difficult to interpret statement does:

```
#2 a <= repeat(2) @(posedge clk) d;
```

11.25 (Fork–join, 20 min.) Write some test code to compare the behavior of the code fragments shown in Table 11.19.

TABLE 11.19 Fork-and-join examples for Problem 11.25.

	(a)	(b)	(c)	(d)

Code fragment	```fork```
	```a = b;```
	```b = a;```
	```join```
		```fork```
		```a <= b;```
		```b <= a;```
		```join```
			```fork```
			```#1 a = b;```
			```#1 b = a;```
			```join```
				```fork```
				```a = #1 b;```
				```b = #1 a;```
				```join```

11.26 (Blocking and nonblocking assignments, 20 min.) Simulate the following code and explain the results:

```
module nonblocking; reg Y;
    always begin Y <= #10 1;Y <= #20 0;#10; end
    always begin $display($time,,"Y=",Y); #10; end
```

```
initial #100 $finish;
endmodule
```

11.27 (*Flip-flop code, 10 min.) Explain why this flip-flop does not work:

```
module Dff_Res_Bad(D,Q,Clock,Reset);
output Q; input D,Clock,Reset; reg Q; wire D;
always @(posedge Clock) if (Reset !== 1) Q = D; always if (Reset == 1)
Q = 0;
end endmodule
```

11.28 (D flip-flop, 10 min.) Test the following D flip-flop model:

```
module DFF (D, Q, Clk, Rst);
parameter width = 1,  reset_value = 0;
input [width-1:0] D; output [width-1:0] Q; reg [width-1:0] Q;
input Clk,Rst;
initial Q = {width{1'bx}};
always @ ( posedge Clk or negedge Rst )
if ( Rst == 0 ) Q <= #1 reset_value; else Q <= #1 D;
endmodule
```

11.29 (D flip-flop with scan, 10 min.) Explain the following model:

```
module DFFSCAN (D, Q, Clk, Rst, ScEn, ScIn, ScOut);
parameter width = 1,  reset_value = 0;
input [width-1:0] D; output [width-1:0] Q; reg [width-1:0] Q;
input Clk,Rst,ScEn,ScIn; output ScOut;
initial Q = {width{1'bx}};
always @ ( posedge Clk or negedge Rst ) begin
   if ( Rst == 0 )      Q <= #1 reset_value;
   else if (ScEn)       Q <= #1 {Q,ScIn};
   else                 Q <= #1 D;
end
assign ScOut=Q[width-1];
endmodule
```

11.30 (Pads, 30 min.) Test the following model for a bidirectional I/O pad:

```
module PadBidir (C, Pad, I, Oen); // active low enable
parameter width=1, pinNumbers="", \strength =1, level="CMOS",
pull="none", externalVdd=5;
output [width-1:0] C; inout [width-1:0] Pad; input [width-1:0] I;
input Oen;
assign #1 Pad = Oen ? {width{1'bz}} : I;
assign #1 C = Pad;
endmodule
```

Construct and test a model for a three-state pad from the above.

11.31 (Loops, 15 min.) Explain and correct the problem in the following code:

```
module Loop_Bad; reg [3:0] i; reg [31:0] DBus;
initial DBus = 0;
```

```
initial begin #1; for (i=0; i<=15; i=i+1) DBus[i]=1; end
initial begin
$display("DBus = %b",DBus); #2; $display("DBus = %b",DBus); $stop;
end endmodule
```

11.32 (Arithmetic, 10 min.) Explain the following:

```
integer IntA;
IntA = -12 / 3; // result is -4
IntA = -'d 12 / 3; // result is 1431655761
```

Determine and explain the values of intA and regA after each assignment statement in the following code:

```
integer intA; reg [15:0] regA;
intA = -4'd12; regA = intA/3; regA = -4'd12;
intA = regA/3; intA = -4'd12/3; regA = -12/3;
```

11.33 (Arithmetic overflow, 30 min.) Consider the following:

```
reg [7:0] a, b, sum; sum = (a + b) >> 1;
```

The intent is to add a and b, which may cause an overflow, and then shift sum to keep the carry bit. However, because all operands in the expression are of an 8-bit width, the expression (a + b) is only 8 bits wide, and we lose the carry bit before the shift. One solution forces the expression (a + b) to use at least 9 bits. For example, adding an integer value of 0 to the expression will cause the evaluation to be performed using the bit size of integers [LRM 4.4.2]. Check to see if the following alternatives produce the intended result:

```
sum = (a + b + 0) >> 1;

sum = {0,a} + {0,b} >> 1;
```

11.34 (*Data slip, 60 min.) Table 11.20 shows several different ways to model the connection of a 2-bit shift register. Determine which of these models suffer from data slip. In each case show your simulation results.

11.35 (**Timing, 30 min.) What does a simulator display for the following?

```
assign p = q; initial begin q = 0; #1 q = 1; $display(p); end
```

What is the problem here? Conduct some experiments to illustrate your answer.

11.36 (Port connections, 10 min.) Explain the following declaration:

```
module test (.a(c), .b(c));
```

11.37 (**Functions and tasks, 30 min.) Experiment to determine whether invocation of a function (or task) behaves as a blocking or nonblocking assignment.

11.38 (Nonblocking assignments, 10 min.) Predict the output of the following model:

```
module e1; reg a, b, c;
initial begin a = 0; b = 1; c = 0; end
always c = #5 ~c; always @(posedge c) begin a <= b; b <= a; end
endmodule
```

TABLE 11.20 Data slip (Problem 11.34).

	Alternative	Data slip?
1	`always @(posedge Clk) begin Q2 = Q1; Q1 = D1; end`	
2	`always @(posedge Clk) begin Q1 = D1; Q2 = Q1; end`	
3	`always @(posedge Clk) begin Q1 <= #1 D1; Q2 <= #1 Q1; end`	
4	`always @(posedge Clk) Q1 = D1; always @(posedge Clk) Q2 = Q1;`	Y
5	`always @(posedge Clk) Q1 = #1 D1; always @(posedge Clk) Q2 = #1 Q1;`	N
6	`always @(posedge Clk) #1 Q1 = D1; always @(posedge Clk) #1 Q2 = Q1;`	
7	`always @(posedge Clk) Q1 <= D1; always @(posedge Clk) Q2 <= Q1;`	
8	`module FF_1 (Clk, D1, Q1); always @(posedge Clk) Q1 = D1; endmodule` `module FF_2 (Clk, Q1, Q2); always @(posedge Clk) Q2 = Q1; endmodule`	
9	`module FF_1 (Clk, D1, Q1); always @(posedge Clk) Q1 <= D1; endmodule` `module FF_2 (Clk, Q1, Q2); always @(posedge Clk) Q2 <= Q1; endmodule`	

11.39 (Assignment timing, 20 min.) Predict the output of the following module and explain the timing of the assignments:

```
module e2; reg a, b, c, d, e, f;
initial begin a = #10 1; b = #2 0; c = #4 1; end
initial begin d <= #10 1; e <= #2 0; f <= #4 1; end
endmodule
```

11.40 (Swap, 10 min.) Explain carefully what happens in the following code:

```
module e3; reg a, b;
initial begin a = 0; b = 1; a <= b; b <= a; end
endmodule
```

11.41 (*Overwriting, 30 min.) Explain the problem in the following code, determine what happens, and conduct some experiments to explore the problem further:

```
module m1; reg a;
initial a = 1;
initial begin a <= #4 0; a <= #4 1; end
endmodule
```

11.42 (*Multiple assignments, 30 min.) Explain what happens in the following:

```
module m2; reg r1; reg [2:0] i;
initial begin
r1 = 0; for (i = 0; i <= 5; i = i+1) r1 <= # (i*10) i[0]; end
endmodule
```

11.43 (Timing, 30 min) Write a model to mimic the behavior of a traffic light signal. The clock input is 1 MHz. You are to drive the lights as follows (times that the lights are on are shown in parentheses): green (60 s), yellow (1 s), red (60 s).

11.44 (Port declarations, 30 min.) The rules for port declarations are as follows: "The port expression in the port definition can be one of the following:

- a simple identifier
- a bit-select of a vector declared within the module
- a part-select of a vector declared within the module
- a concatenation of any of the above

Each port listed in the module definition's list of ports shall be declared in the body of the module as an input, output, or inout (bidirectional). This is in addition to any other declaration for a particular port—for example, a reg, or wire. A port can be declared in both a port declaration and a net or register declaration. If a port is declared as a vector, the range specification between the two declarations of a port shall be identical" [Verilog LRM 12.3.2].

Compile the following and comment (you may be surprised at the results):

```
module stop (); initial #1 $finish; endmodule
module Outs_1 (a); output [3:0] a; reg [3:0] a;
initial a <= 4'b10xz; endmodule
module Outs_2 (a); output [2:0] a; reg [3:0] a;
initial a <= 4'b10xz; endmodule
module Outs_3 (a); output [3:0] a; reg [2:0] a;
initial a <= 4'b10xz; endmodule
module Outs_4 (a); output [2:0] a; reg [2:0] a;
initial a <= 4'b10xz; endmodule
module Outs_5 (a); output a; reg [3:0] a;
initial a <= 4'b10xz; endmodule
module Outs_6 (a[2:0]); output [3:0] a; reg [3:0] a;
initial a <= 4'b10xz; endmodule
module Outs_7 (a[1]); output [3:0] a; reg [3:0] a;
initial a <= 4'b10xz; endmodule
module Outs_8 (a[1]); output a; reg [3:0] a;
always a <= 4'b10xz; endmodule
```

11.45 (Specify blocks, 30 min.)

a. Describe the pin-to-pin timing of the following module. Build a testbench to demonstrate your explanation.

```
module XOR_spec (a, b, z); input a, b: output z; xor x1 (z, a, b);
specify
   specparam tnr = 1, tnf = 2 specparam tir = 3, tif = 4;
   if ( a)(b => z) = (tir, tif); if ( b)(a => z) = (tir, tif);
   if (~a)(b => z) = (tnr, tnf); if (~b)(a => z) = (tnr, tnf);
endspecify
endmodule
```

b. Write and test a module for a 2:1 MUX with inputs A0, A1, and sel; output z; and the following delays: A0 to z: 0.3ns (rise) and 0.4ns (fall); A1 to z: 0.2ns (rise) and 0.3 ns (fall); sel to z=0.5 ns.

11.46 (Design contest, **60 min.) In 1995 John Cooley organized a contest between VHDL and Verilog for ASIC designers. The goal was to design the fastest 9-bit counter in under one hour using Synopsys synthesis tools and an LSI Logic vendor technology library. The Verilog interface is as follows:

```
module counter (data_in, up, down, clock,
   count_out, carry_out, borrow_out, parity_out);
output [8:0]  count_out;
output carry_out, borrow_out, parity_out;
input  [8:0]  data_in; input clock, up, down;
reg    [8:0]  count_out; reg carry_out,  borrow_out, parity_out;
// Insert your design here.
endmodule
```

The counter is positive-edge triggered, counts up with up='1' and down with down='1'. The contestants had the advantage of a predefined testbench with a set of test vectors; you do not. Design a model for the counter and a testbench.

11.47 (Timing checks, ***60 min.+) Flip-flops with preset and clear require more complex timing-check constructs than those described in Section 11.13.3. The following BNF defines a **controlled timing-check event**:

```
controlled_timing_check_event ::= timing_check_event_control
specify_terminal_descriptor [ &&& timing_check_condition ]

timing_check_condition ::=
  scalar_expression | ~scalar_expression
| scalar_expression == scalar_constant
| scalar_expression === scalar_constant
| scalar_expression != scalar_constant
| scalar_expression !== scalar_constant
```

The scalar expression that forms the conditioning signal must be a scalar net, or else the least significant bit of a vector net or a multibit expression value is used. The comparisons in the timing check condition may be **deterministic** (using ===, !==, ~, or no operator) or **nondeterministic** (using == or !=). For deterministic comparisons, an 'x' result disables the timing check. For nondeterministic comparisons, an 'x' result enables the timing check.

As an example the following **unconditioned timing check**,

```
$setup(data, posedge clock, 10);
```

performs a setup timing check on every positive edge of clock, as was explained in Section 11.13.3. The following controlled timing check is enabled only when clear is high, which is what is required in a flip-flop model, for example.

```
$setup(data, posedge clock &&& clear, 10);
```

The next example shows two alternative ways to enable a timing check only when clear is low. The second method uses a nondeterministic operator.

```
$setup(data,posedge clock &&&(~clear),10); // clear=x disables check
$setup(data,posedge clock &&&(clear==0),10); // clear=x enables check
```

To perform the setup check only when `clear` and `preset` signals are high, you can add a gate outside the specify block, as follows:

```
and g1(clear_and_preset, clear, set);
```

A controlled timing check event can then use this `clear_and_preset` signal:

```
$setup(data, posedge clock && clear_and_preset, 10);
```

Use the preceding techniques to expand the D flip-flop model, `dff_udp`, from Section 11.13.3 to include asynchronous active-low preset and clear signals as well as an output, `qbar`. Use the following module interface:

```
module dff(q, qbar, clock, data, preset, clear);
```

11.48 (Verilog BNF, 30 min.) Here is the "old" BNF definition of a sequential block (used in the Verilog reference manuals and the OVI LRM). Are there any differences from the "new" version?

```
<sequential_block> ::=
   begin <statement>* end
   or
   begin: <block_IDENTIFIER> <block_declaration>*
     <statement>*
   end

<block_declaration> ::= parameter <list_of_param_assignment>;
   or reg <range>? <attribute_decl>*
     <list_of_register_variable>;
   or integer <attribute_decl>* <list_of_register_variable>;
   or real <attribute_decl>* <list_of_variable_IDENTIFIER>;
   or time <attribute_decl>* <list_of_register_variable>;
   or event <attribute_decl>* <list_of_event_IDENTIFIER>;

<statement> ::=
   <blocking_assignment>;
   or <non-blocking_assignment>;
   or if(<expression>) <statement_or_null>
     <else <statement_or_null> >?
   or <case or casez or casex>
     (<expression>) <case item>+ endcase
   or forever <statement>
   or repeat(<expression>) <statement>
   or while(<expression>) <statement>
   or for(<assignment>;
     <expression>; <assignment>) <statement>
   or wait(<expression>) <statement_or_null>
   or disable <task_IDENTIFIER>;
   or disable <block_IDENTIFIER>;
   or force <assignment>; or release <value>;
   or <timing_control> <statement_or_null>
   or -> <event_IDENTIFIER>;
```

```
or <sequential_block> or <parallel_block>
or <task_enable> or <system_task_enable>
```

11.49 (Conditional compiler directives, 30 min.) The conditional compiler directives: `define, `ifdef, `else, `endif, and `undef; work much as in C. Write and compile a module that models an AND gate as 'z = a&b' if the variable behavioral is defined. If behavioral is not defined, then model the AND gate as 'and al (z, a, b)'.

11.50 (*Macros, 30 min.) According to the IEEE Verilog LRM [16.3.1] you can create a **macro** with parameters using `define, as the following example illustrates. This is a particularly difficult area of compliance. Does your software allow the following? You may have to experiment considerably to get this to work. *Hint:* Check to see if your software is substituting for the macro text literally or if it does in fact substitute for parameters.

```
`define M_MAX(a, b)((a) > (b) ? (a) : (b))
`define M_ADD(a,b) (a+b)
module macro;
reg m1, m2, m3, s0, s1;
`define var_nand(delay) nand #delay
`var_nand (2) g121 (q21, n10, n11);
`var_nand (3) g122 (q22, n10, n11);
initial begin s0=0; s1=1;
m1 = `M_MAX (s0, s1); m2 = `M_ADD (s0,s1); m3 = s0 > s1 ? s0 : s1;
end
initial #1 $display(" m1=",m1," m2=",m2," m3=",m3);
endmodule
```

11.51 (**Verilog hazards, 30 min.) The MTI simulator, VSIM, is capable of detecting the following kinds of Verilog hazards:

1. WRITE/WRITE: Two processes writing to the same variable at the same time.

2. READ/WRITE: One process reading a variable at the same time it is being written to by another process. VSIM calls this a READ/WRITE hazard if it executed the read first.

3. WRITE/READ: Same as a READ/WRITE hazard except that VSIM executed the write first.

For example, the following log shows how to simulate Verilog code in hazard mode for the example in Section 11.6.2:

```
> vlib work
> vlog -hazards data_slip_1.v
> vsim -c -hazards data_slip_1
...(lines omitted)...
# 100 0    1 1   x
# ** Error: Write/Read hazard detected on Q1 (ALWAYS 3 followed by
ALWAYS 4)
#    Time: 150 ns  Iteration: 1  Instance:/
```

```
# 150 1   1 1  1
...(lines omitted)...
```

There are a total of five hazards in the module `data_slip_1`, four are on `Q1`, but there is another. If you correct the code as suggested in Section 11.6.2 and run VSIM, you will find this fifth hazard. If you do not have access to MTI's simulator, can you spot this additional read/write hazard? *Hint*: It occurs at time zero on `Clk`. Explain.

11.15.1 The Viterbi Decoder

11.52 (Understanding, 20 min.) Calculate the values shown in Table 11.8 if we use 4 bits for the distance measures instead of 3.

11.53 (Testbenches)

a. (30 min.) Write a testbench for the encoder, `viterbi_encode`, in Section 11.12 and reproduce the results of Table 11.7.

b. (30 min.) Write a testbench for the receiver front-end `viterbi_distances` and reproduce the results of Table 11.9 (you can write this stand-alone or use the answer to part a to generate the input). *Hint:* You will need a model for a D flip-flop. The sequence of results is more important than the exact timing. If you do have timing differences, explain them carefully.

11.54 (Things go wrong, 60 min.) Things do not always go as smoothly as the examples in this book might indicate. Suppose you accidentally invert the sense of the reset for the D flip-flops in the encoder. Simulate the output of the faulty encoder with an input sequence $X_n = 0, 1, 2, 3, \dots$ (in other words run the encoder with the flip-flops being reset continually). The output sequence looks reasonable (you should find that it is $Y_n = 0, 2, 4, 6, \dots$). Explain this result using the state diagram of Figure 11.3. If you had constructed a testbench for the entire decoder and did not check the intermediate signals against expected values you would probably never find this error.

11.55 (Subset decoder) Table 11.21 shows the inputs and outputs from the first-stage of the Viterbi decoder, the subset decoder. Calculate the expected output and then confirm your predictions using simulation.

TABLE 11.21   Subset decoder (Problem 11.55).

input	in0	in1	in2	in3	in4	in5	in6	in7	s0	s1	s2	s3	sout0	sout1	sout2	sout3
5	6	7	6	4	1	0	1	4	1	0	1	4				
4	7	6	4	1	0	1	4	6	0	1	4	1				
1	1	0	1	4	6	7	6	4	1	0	1	4				
0	0	1	4	6	7	6	4	1	0	1	4	1				

11.16 Bibliography

The IEEE Verilog LRM [1995] is less intimidating than the IEEE VHDL LRM, because it is based on the OVI LRM, which in turn was based on the Verilog-XL simulator reference manual. Thus it has more of a "User's Guide" flavor and is required reading for serious Verilog users. It is the only source for detailed information on the PLI.

Phil Moorby was one of the original architects of the Verilog language. The Thomas and Moorby text is a good introduction to Verilog [1991]. The code examples from this book can be obtained from the World Wide Web. Palnitkar's book includes an example of the use of the PLI routines [1996].

Open Verilog International (OVI) has a Web site maintained by Chronologic (http://www.chronologic.com/ovi) with membership information and addresses and an ftp site maintained by META-Software (ftp://ftp.metasw.com in /pub/OVI/). OVI sells reference material, including proceedings from the International Verilog HDL Conference.

The newsgroup comp.lang.verilog (with a FAQ—frequently asked questions) is accessible from a number of online sources. The FAQ includes a list of reference materials and book reviews. Cray Research maintained an archive for comp.lang.verilog going back to 1993 but this was lost in January 1997 and is still currently unavailable. Cadence has a discussion group at talkverilog@cadence.com. Wellspring Solutions offers VeriWell, a no-cost, limited capability, Verilog simulator for UNIX, PC, and Macintosh platforms.

There is a free, "copylefted" Verilog simulator, vbs, written by Jimen Ching and Lay Hoon Tho as part of their Master's theses at the University of Hawaii, which is part of the comp.lang.verilog archive. The package includes explanations of the mechanics of a digital event-driven simulator, including event queues and time wheels.

More technical references are included as part of Appendix B.

11.17 References

IEEE Std 1364-95, Verilog LRM. 1995. The Institute of Electrical and Electronics Engineers. Available from The Institute of Electrical and Electronics Engineers, Inc., 345 East 47th Street, New York, NY 10017 USA. [cited on p. 479]

Palnitkar, S. 1996. *Verilog HDL: A Guide to Digital Design and Synthesis.* Upper Saddle River, NJ: Prentice-Hall, 396 p. ISBN 0-13-451675-3.

Thomas, D. E., and P. Moorby. 1991. *The Verilog Hardware Description Language.* 1st ed. Dordrecht, Netherlands: Kluwer, 223 p. ISBN 0-7923-9126-8, TK7885.7.T48 (1st ed.). ISBN 0-7923-9523-9 (2nd ed.).